21世纪高等学校计算机类课程创新规划教材·微课版

Android
应用程序设计（修订版）

◎ 张思民　编著

清华大学出版社
北京

内 容 简 介

本书是面向Android系统的初学者的入门教程，内容几乎涵盖了Android相关的所有技术。本书大致可以分成两个部分，第一部分（第1~4章）主要介绍Android SDK开发环境的安装、应用程序的结构、用户界面的组件及其设计方法，第二部分（第5~10章）主要介绍较高级的主题，内容包括异常处理与多线程、后台服务与系统服务技术、图形与多媒体处理技术、数据库技术及输入/输出流的处理技术、网络通信技术、地图服务及传感器检测技术等。

本书在叙述上浅显易懂，对每一个知识点都配了相应的例题。本书提供了所有例题的源代码、电子课件，以及本书大部分例题的视频教学演示（扫描二维码）。

本书可以作为高等院校及各类培训学校Android系统课程的教材，也可以作为学习Android程序设计的编程人员的自学用书。

本书封面贴有清华大学出版社防伪标签，无标签者不得销售。

版权所有，侵权必究。举报：010-62782989，beiqinquan@tup.tsinghua.edu.cn。

图书在版编目（CIP）数据

Android应用程序设计 / 张思民编著. —修订版. —北京：清华大学出版社，2018（2022.8重印）
（21世纪高等学校计算机类课程创新规划教材·微课版）
ISBN 978-7-302-49578-9

Ⅰ.①A… Ⅱ.①张… Ⅲ.①移动终端－应用程序－程序设计 Ⅳ.①TN929.53

中国版本图书馆CIP数据核字（2018）第027488号

责任编辑：魏江江
封面设计：刘　键
责任校对：徐俊伟
责任印制：杨　艳

出版发行：清华大学出版社
　　网　　址：http://www.tup.com.cn，http://www.wqbook.com
　　地　　址：北京清华大学学研大厦A座　　邮　编：100084
　　社 总 机：010-83470000　　邮　购：010-62786544
　　投稿与读者服务：010-62776969，c-service@tup.tsinghua.edu.cn
　　质量反馈：010-62772015，zhiliang@tup.tsinghua.edu.cn
　　课件下载：http://www.tup.com.cn，010-83470236
印　刷　者：北京富博印刷有限公司
装　订　者：北京市密云县京文制本装订厂
经　　　销：全国新华书店
开　　　本：185mm×260mm　　印　张：19.75　　字　数：481千字
版　　　次：2013年3月第1版　2018年5月第2版　印　次：2022年8月第6次印刷
印　　　数：8501~9500
定　　　价：39.50元

产品编号：078298-01

前 言

Android 系统自 2007 年推出以来，应用越来越广泛。除了手机、平板电脑使用 Android 系统之外，其他嵌入式系统也大量使用 Android 系统来设计。例如，车载设备、医疗设备、VoIP 电话和智能电视等厂商纷纷推出 Android 系统产品。可以说，Android 系统如日中天，相信将来还会有更好的发展。

1．本书特点

作为一本教材，本书有以下特点。

（1）易学易懂。本书面向 Android 系统的初学者，在叙述方式上浅显易懂，摒弃枯燥的理论，尽可能使用图示加以说明。对每一个知识点，都配了相应的例题。所有例题均短小精悍，适合课堂教学讲授。读者学完每一章内容后都可以编写出相应功能的程序。

（2）解释详细。对每一个例题，均进行了详细分析和解释，既可以帮助读者学习理解知识和概念，大大降低学习难度，又具有启发性。

（3）Java 语言零基础学习。为了帮助没有 Java 语言基础的读者学习 Android 系统，特别安排了一章介绍 Java 基础知识的内容。

（4）配有视频教学演示。书中大部分例题均录制了教学视频，详细地记录了设计的操作过程，帮助读者更加轻松、迅速地理解和掌握本书内容。

2．学习方法

学习 Android 程序设计，应该循序渐进、由浅入深，不能跳跃式地进行，前面的内容还没搞清楚，就急于学习后面的内容，这样只会事倍功半，欲速则不达。

应该说，学习任何一种编程技术都会有一定难度。因此，要强调动手实践，多编程、多练习，熟能生巧，从学习中体验到程序设计的乐趣和成功的喜悦，增强学习信心。

3．本书内容

本书在内容结构上大致可以分成两个部分。

第一部分（第 1～4 章）主要介绍 Android SDK 开发环境的安装、应用程序的结构、用户界面的组件及其设计方法，该部分内容是学习 Android 程序设计的入门基础。

第 1 章主要讲解 Android SDK 开发环境的安装，并说明如何下载 Android SDK 和如何从头开始创建新的应用程序。第 2 章简要介绍 Java 语言基础知识，为不熟悉 Java 语言的读者提供帮助，对于已有 Java 语言基础的读者，可以跳过本章。第 3～4 章讲解如何使用布局和视图创建用户界面，介绍了用户图形界面的常用组件及多用户界面程序的开发。

第二部分（第 5～10 章）主要介绍较高级的主题，内容包括异常处理及多线程、图形与多媒体处理技术、后台服务与系统服务技术、数据库技术及输入/输出流的处理技术、网络通信技术、地图服务及传感器检测技术等。

第 5 章讲解 Android 的异常处理方法以及多线程。第 6 章讲解图形与多媒体处理技术，介绍了绘制几何图形的基本方法、处理触摸屏事件的方法，还详细讨论了音频播放和视频播放的设计，以及录音、照相和文本转换语音技术，最后详细讲解了如何处理图像的缩放、变形、颜色等数字图像处理技术。第 7 章讲解后台服务与系统服务，以及系统功能调用。第 8 章讲解数据存储技术，介绍了 SQLite 数据库存储方式、文件存储方式和 XML 文件的 SharedPreferences 存储方式。第 9 章讲解网络通信，介绍了 Socket 套接字编程、基于 Web 编程和与 JavaScript 脚本交互的编程技术，以及无线网络通信技术 WiFi 的程序设计方法。第 10 章讲解地图服务及传感器检测技术，地图服务主要介绍地图查询和贴图的方法，传感器检测主要介绍重力加速度的应用。

书中所有例题均已在 Eclipse + ADT 环境下运行通过。本书提供了所有例题的源代码、电子课件。

参加本书编写、校对及程序测试工作的还有梁维娜、张静文、杨军民、颜敏敏等，在此表示感谢。

由于编者水平有限，书中难免有不足之处，敬请读者批评指正。

编　者

2018 年 1 月

目 录

第 1 章 **Android 系统及其开发过程** ·· 1
1.1 Android 系统概述 ··· 1
1.2 安装 Android SDK 开发环境 ··· 2
    1.2.1 安装 Android SDK 前必要的准备 ·· 2
    1.2.2 安装 Android SDK 详解 ·· 3
    1.2.3 设置环境变量 ··· 7
1.3 Android API 和在线帮助文档 ·· 7
1.4 Android 应用程序的开发过程 ··· 8
    1.4.1 开发 Android 应用程序的一般过程 ··· 8
    1.4.2 生成 Android 应用程序框架 ·· 9
    1.4.3 编写 MainActivity.java ·· 10
    1.4.4 配置应用程序的运行参数 ·· 11
    1.4.5 在模拟器中运行应用程序 ·· 12
1.5 Android 应用程序结构 ··· 12
    1.5.1 目录结构 ·· 12
    1.5.2 Android 应用程序架构分析 ·· 18
1.6 Android 应用程序设计示例 ··· 19
习题 1 ··· 21
第 2 章 **Java 语法概述** ·· 22
2.1 语法基础 ·· 22
    2.1.1 数据类型 ·· 22
    2.1.2 常量与变量 ·· 23
    2.1.3 对变量赋值 ·· 24
    2.1.4 关键字 ··· 24
    2.1.5 转义符 ··· 24
2.2 基本数据类型应用示例 ·· 25
    2.2.1 整型与浮点型 ··· 25
    2.2.2 字符型 ··· 27
    2.2.3 布尔型 ··· 28
    2.2.4 数据类型的转换 ·· 29
2.3 程序控制语句 ··· 30
    2.3.1 语句的分类 ·· 30

		2.3.2　顺序控制语句 31
		2.3.3　if 语句 32
		2.3.4　switch 语句 34
		2.3.5　循环语句 35
		2.3.6　转语句 40
	2.4　类与对象 42
		2.4.1　类的定义 42
		2.4.2　对象 44
		2.4.3　接口 46
		2.4.4　包 47
	2.5　XML 语法简介 47
	习题 2 51

第 3 章　Android 用户界面设计 53
	3.1　用户界面组件包 widget 和 View 类 53
	3.2　文本标签与按钮 54
		3.2.1　文本标签 54
		3.2.2　按钮 56
	3.3　文本编辑框 61
	3.4　Android 布局管理 64
		3.4.1　布局文件的规范与重要属性 64
		3.4.2　常见的布局方式 65
	3.5　进度条和选项按钮 72
		3.5.1　进度条 72
		3.5.2　选项按钮 74
	3.6　图像显示与画廊组件 80
		3.6.1　图像显示 ImageView 类 80
		3.6.2　画廊组件 Gallery 与图片切换器 ImageSwitcher 84
	3.7　消息提示 88
	3.8　列表组件 91
		3.8.1　列表组件 ListView 类 91
		3.8.2　列表组件 ListActivity 类 94
	3.9　滑动抽屉组件 96
	习题 3 100

第 4 章　多个用户界面的程序设计 102
	4.1　页面切换与传递参数值 102
		4.1.1　传递参数组件 Intent 102
		4.1.2　Activity 页面切换 102
		4.1.3　应用 Intent 在 Activity 页面之间传递数据 106
	4.2　菜单 110

		4.2.1 选项菜单	110
		4.2.2 上下文菜单	112
	4.3	对话框	114
		4.3.1 消息对话框	114
		4.3.2 其他几种常用对话框	120
	习题 4		123
第 5 章	异常处理与多线程		124
	5.1	异常处理	124
	5.2	多线程	126
		5.2.1 线程与多线程	126
		5.2.2 线程的生命周期	127
		5.2.3 线程的数据通信	128
		5.2.4 创建线程	130
	习题 5		137
第 6 章	图形与多媒体处理		138
	6.1	绘制几何图形	138
		6.1.1 几何图形绘制类	138
		6.1.2 几何图形的绘制过程	139
	6.2	触摸屏事件处理	144
		6.2.1 简单触摸屏事件	144
		6.2.2 手势识别事件	150
	6.3	音频播放	153
		6.3.1 多媒体处理包	153
		6.3.2 媒体处理播放器	154
		6.3.3 播放音频文件	155
	6.4	视频播放	161
		6.4.1 应用媒体播放器播放视频	161
		6.4.2 应用视频视图播放视频	164
	6.5	录音与拍照	166
		6.5.1 用于录音、录像的 MediaRecorder 类	166
		6.5.2 录音示例	167
		6.5.3 拍照	170
	6.6	将文本转换成语音	176
	6.7	图像处理技术	178
		6.7.1 处理图像的颜色矩阵	178
		6.7.2 处理图像的坐标变换矩阵	184
	习题 6		192
第 7 章	后台服务与系统服务技术		193
	7.1	后台服务 Service	193

7.2 信息广播机制 Broadcast ... 197
7.3 系统服务 ... 206
　7.3.1 Android 的系统服务 ... 206
　7.3.2 系统通知服务 Notification ... 206
　7.3.3 系统定时服务 AlarmManager ... 209
　7.3.4 系统功能的调用 ... 212
习题 7 ... 215

第 8 章 数据存储 ... 216
8.1 SQLite 数据库 ... 216
　8.1.1 SQLite 数据库简介 ... 216
　8.1.2 管理和操作 SQLite 数据库的对象 ... 217
　8.1.3 SQLite 数据库的操作命令 ... 218
8.2 文件处理 ... 230
　8.2.1 输入流和输出流 ... 230
　8.2.2 处理文件流 ... 231
8.3 轻量级存储 SharedPreferences ... 238
习题 8 ... 240

第 9 章 网络通信 ... 241
9.1 网络编程的基础知识 ... 241
　9.1.1 IP 地址和端口号 ... 241
　9.1.2 套接字 ... 244
9.2 基于 TCP 的网络程序设计 ... 246
9.3 基于 HTTP 的网络程序设计 ... 251
9.4 Web 视图 ... 255
　9.4.1 浏览器引擎 WebKit ... 255
　9.4.2 Web 视图对象 ... 255
　9.4.3 调用 JavaScript ... 258
9.5 无线网络通信技术 WiFi ... 266
习题 9 ... 272

第 10 章 地图服务及传感器检测技术 ... 273
10.1 Google 地图 ... 273
　10.1.1 Google Maps 包 ... 273
　10.1.2 导入 Google 地图 API 的 Maps 包 ... 274
　10.1.3 显示地图 MapView 类 ... 274
　10.1.4 添加 Google 地图的贴图 ... 279
10.2 位置服务 ... 282
10.3 传感器检测技术 ... 286
　10.3.1 传感器简介 ... 286

    10.3.2 加速度传感器的应用示例 ································· 289

 习题 10 ··· 297

附录 A **JavaSDK 及 Eclipse 的安装与配置** ························· 298

附录 B **Android 的调试工具** ····································· 300

附录 C **Map API Key 的申请过程** ································ 303

第 1 章　Android 系统及其开发过程

1.1　Android 系统概述

2007 年 11 月 5 日，Google 公司推出了基于 Linux 操作系统的智能手机平台 Android 系统。Android 系统由操作系统、中间件、用户界面程序和应用软件等组成。为了推动 Android 系统发展，Google 与手机制造商、电信运营商、半导体公司、软件公司等 65 家企业联手组成了一个商业联盟组织——OHA（Open Handset Alliance，开放手机联盟），制定了基于 Android 的移动设备生产和开发标准。

Android 的出现绝非偶然，它是由传统的移动电话系统开发模式演变而来的一种符合时代潮流的新型移动开发模式的产物。Android 的出现为移动开发者带来了新的机遇与挑战。

移动电话的开发经历了传统移动电话的开发、半开放式移动电话的开发、全开放式移动电话的开发 3 个发展阶段。

（1）传统移动电话的开发：移动电话厂商制作移动电话出售，厂商有自己的研发机构，也依靠其他公司提供的解决方案来完成移动电话的开发工作。通俗点说，就是买了移动电话，里面的功能已经固定，没有扩展的机会。

（2）半开放式移动电话的开发：随着自定义需求的增加，移动开发走向了半开放模式。在这种模式下，厂商制造移动电话出售，预置了部分基本软件功能，但是支持增加第三方应用程序，用户可以根据自己的需要选择下载安装。在这种模式下，第三方应用程序开发接口是开放的，但是系统本身是不开放的，因此只能称为半开放模式。

（3）全开放式移动电话的开发：Android 的出现，是全开放开发模式的缩影，不仅第三方应用程序接口开放，Android 系统本身也是完全开放的。各个厂商在统一的平台上开发移动电话，第三方开发移动应用。如果系统不能满足需求，则可以在系统中增加新的功能，这就是全开放的优势。

移动电话经过 20 年的发展，已经不仅仅是一个移动的通信工具，随着 3G 技术的发展，移动电话正向着智能化的方向迈进，移动电话正逐渐成为多种应用工具的功能载体，如通信工具、网络工具、媒体播放器、媒体获取设备、多类型的连接设备、信息感知终端、视频电话、个性化定制平台等。

Android 系统诞生在开放时代的背景下，其全开放的智能移动平台、多硬件平台的支持、使用众多标准化的技术、核心技术完整、完善的 SDK 和文档、完善的辅助开发工具等特点与智能手机的发展方向紧密相连，它将代表并引领新时代的技术潮流。

对于开发者而言，Android 开发分为两大类：

（1）移植开发移动电话系统。移植开发是为了将 Android 系统运行在手持式移动设备上，在具体的硬件系统上构建 Android 软件系统。这种类型的开发在 Andriod 底层进行，需要移植开发 Linux 中相关的设备驱动程序及 Android 本地框架中的硬件抽象层，也就是需要将设备驱动与 Android 系统联系起来。Android 系统对硬件抽象层都有标准的接口定义，移植时，只需实现这些接口即可。

（2）Android 应用程序开发。应用程序开发可以基于硬件设备（用于测试的实体手机），也可以基于 Android 模拟器。应用开发处于 Android 系统的顶层，使用 Android 系统提供的 Java 框架（API）进行开发设计工作，是大多数开发者从事的开发工作。本书所介绍的 Android 应用程序设计，都是在这个层次上进行的。

1.2　安装 Android SDK 开发环境

视频演示

1.2.1　安装 Android SDK 前必要的准备

1. Android 系统开发的操作平台与软件环境要求

目前，对 Android 系统开发支持的操作系统有：

- Windows 7/8/10（32 位或 64 位）
- Mac OS X10 或更高版本
- Linux（GNOME 或 KDE 桌面）

本书讲解主要以 Windows 系统为例，其他操作系统下的 Android 开发请参考 Android 的开发文档。

对于 Android 系统开发的软件环境，在这里主要介绍 Eclipse+ADT（Android Development Tools 插件）。因此，需要安装 Java 8 以上和 Eclipse 3.3 以上版本的环境。有关 Java SDK 以及 Eclipse 的安装与配置请参考附录。

2. 下载最新版本的 Android SDK 软件

这里以 2017 年 4 月 Google 发布的移动操作系统平台的版本 Android 8.0 为例，其面向开发人员的 SDK 版本（android-sdk_r26.0.2）也同步推出（截止到本书校对书稿时，Google 发布的最新版本为 android-sdk_r26.0.2）。

可以到 Android 官方网站 http://developer.android.com/sdk/index.html 下载最新的系统软件，见表 1-1（以版本 Android 8.0 为例）。

表 1-1　下载 Android 系统安装包

安 装 平 台	系统安装包	Size
Windows	android-sdk_r26.0.2-windows.zip	90 379 469 bytes
	installer_r26.0.2-windows.exe（推荐）	70 495 456 bytes
Mac	android-sdk_r26.0.2-macosx.zip	58 218 455 bytes
Linux	android-sdk_r26.0.2-linux.tgz	82 616 305 bytes

1.2.2 安装 Android SDK 详解

1. 运行 Android SDK 安装文件

解压安装包文件 sdk-tools-windows-r26.zip，得到如图 1.1 所示的系统安装框架，然后运行其中的 tools\bin\sdkmanager 来安装 Android SDK。

2. 运行 SDK Manager.exe 文件

在命令行窗口，进入 tools\bin 目录，输入安装命令：

```
sdkmanager "platform-tools" "platforms;android-26"
```

也可以在 Eclipse 中运行 SDK Manager 项，弹出如图 1.2 所示的 SDK 管理窗口，系统自动搜索所有版本的系统安装包。单击位于窗口下方的 Install Packages 按钮，系统自动下载并安装这些选中的包，该过程需要比较长的时间。

图 1.1 Android SDK 系统安装框架

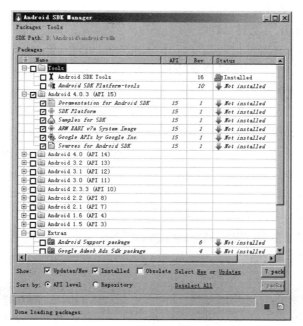

图 1.2 Android SDK 的管理窗口

Android SDK 系统在安装完之后会提示是否安装 ADB（Android Debug Bridge）。ADB 是开发 Android 应用项目的调试工具，这里要确认安装。在 Android SDK 所有系统文件安装完之后，打开安装目录，其目录结构如图 1.3 所示。

目录结构中主要文件夹的作用如下。

- add-ons：放置 Google 提供的 API 包，包括 Google 地图 API 等。
- docs：放置 Android 系统的帮助文档和说明文档。
- platforms：针对每个 SDK 版本提供与其相对应的 API 包。
- tools 和 platform-tools：放置通用的工具文件，如 Android 模拟器 AVD、SQLite 数

据库、调试工具 ADB、创建模拟的 SD 卡工具 mksdcard 等。为了方便地使用这些工具，通常将它们设置为系统环境变量。

图 1.3　安装 Android SDK 系统完毕后的目录结构

- samples：放置每个 SDK 版本提供的示例程序。
- system-images：由于 Android 是基于 Linux 的系统，该文件夹中放置了不同版本的 img 系统映像文件。

3. 安装 Android 开发工具 ADT

Android 为 Eclipse 开发环境提供了一个外挂插件 ADT（Android Development Tools），该插件为用户提供了一个强大的综合环境用于开发 Android 应用程序。ADT 扩展了 Eclipse 的功能，可以让用户快速地建立 Android 项目，创建应用程序界面，在基于 Android 框架 API 的基础上添加组件，以及用 SDK 工具集调试应用程序。

下面详细介绍安装和配置 ADT 的基本方法和步骤。

（1）打开 Eclipse 设置工作目录。在安装 Android 插件 ADT 之前必须已经安装好 Eclipse 集成开发环境，如果尚未安装，可参照附录 A 自行安装。启动 Eclipse，第一次启动时会要求用户设置工作目录，如图 1.4 所示。作者建立了一个 D:\AdroidTest\android-SDK 工作目录，读者可根据自己的需要设置相应的工作目录。

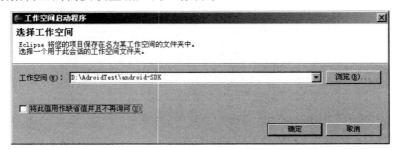

图 1.4　设置 Android 工作目录

（2）安装 ADT 插件。在 Eclipse 中，选择"帮助"（Help）→"安装新软件"（Install New Software）命令，在弹出的 Install 对话框中单击 Add 按钮，然后在 Add Repository 对话框的 Location 文本框中输入 https://dl-ssl.google.com/android/eclipse/，单击"确定"按钮，如图 1.5 所示。

（3）设置 ADT 的首选项。顺利安装 ADT 后，在 Eclipse 中，选择"窗口"（Window）→"首选项"（Preferences）命令，弹出"首选项"对话框，在 SDK Location 文本框中设置安装 Android SDK 的绝对路径，如图 1.6 所示。

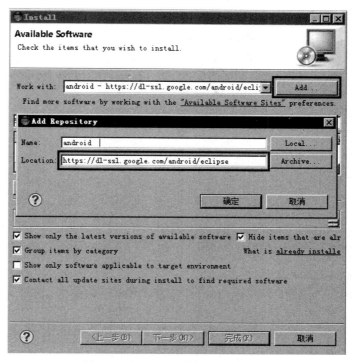

图 1.5　在 Eclipse 中安装 ADT 插件

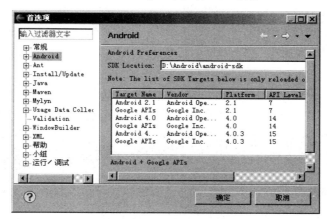

图 1.6　设置 ADT 的首选项

4．创建 Android 虚拟设备 AVD

Android 应用程序可以在实体手机上执行，也可以创建一个 Android 虚拟设备 AVD（Android Virtual Device）来测试。每一个 Android 虚拟设备 AVD 模拟一套虚拟环境来运行 Android 操作系统平台，这个平台有自己的内核、系统图像、外观显示、用户数据区和仿真的 SD 卡等。

下面介绍如何创建一个 Android 虚拟设备 AVD。

Eclipse 集成开发环境提供了 Android SDK and AVD Manager 功能，读者可以用它来创建 Android 虚拟设备 AVD。

（1）选择 Eclipse 中的"窗口"（Window）→AVD Manager 命令，在弹出的 Android Virtual

Device Manager 对话框中可以看到已创建的 AVD。单击右边的 New 按钮创建一个新的 AVD，如图 1.7 所示。

（2）在弹出的 Create new Android Virtual Device（AVD）对话框中，输入或选择如图 1.8 所示的各项内容，单击 Create AVD 按钮，创建一个新的 AVD。

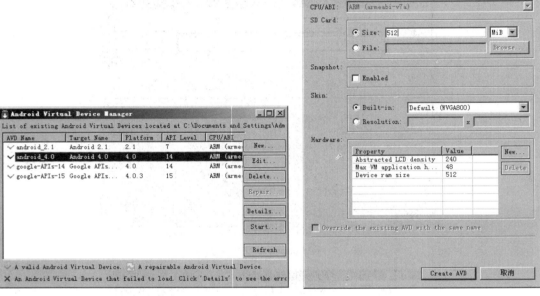

图 1.7　Android Virtual Device Manager 对话框　　　图 1.8　创建新的 AVD

（3）运行 AVD 模拟器。在如图 1.7 所示的 Android Virtual Device Manager 对话框中选择已经建立的 AVD，单击 Start 按钮，可以启动 AVD 模拟器。启动 AVD 模拟器的过程时间很长，建议打开后不要关闭，可以在该模拟器上测试 Android 应用程序。启动的 AVD 模拟器如图 1.9 所示。

图 1.9　Android 的 AVD 模拟器

1.2.3 设置环境变量

安装完 Android SDK 系统后，还要设置环境变量，即把 Android SDK 系统目录下的 platform-tools 的路径设置到系统变量中。右击"我的电脑"，在弹出的快捷菜单中选择"属性"命令，在"系统属性"对话框中选择"高级"选项卡，然后单击"环境变量"按钮，在弹出的"环境变量"对话框的"系统变量"下方找到 Path 变量，单击"编辑"按钮，在"编辑系统变量"对话框的"变量值"文本框中输入 Android SDK 安装目录下的 platform-tools 的完整路径，如图 1.10 所示。

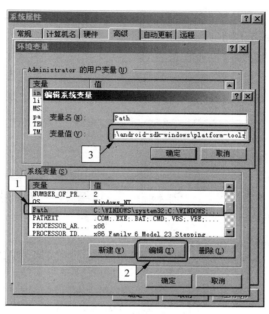

图 1.10　设置 Android 环境变量

1.3　Android API 和在线帮助文档

1. Android API

Android 为用户安装了它所提供的标准类库。所谓标准类库，就是把程序设计所需要的常用方法和接口分类封装成包，Android 所提供的标准类库就是 Android API。

Android 包中封装了程序设计所需要的主要应用类，本书用到了以下十几种包。

- Android.app：封装了高层的程序模型，提供基本的运行环境。
- Android.content：封装了各种对设备上的数据进行访问和发布的类。
- Android.database：通过内容提供者浏览和操作数据库。
- Android.graphics：底层的图形库，包含画布、颜色过滤、点、矩形，可以将它们直接绘制到屏幕上。
- Android.location：封装了定位和相关服务的类。
- Android.media：封装了一些类管理多种音频、视频的媒体接口。
- Android.net：封装了帮助网络访问的类，超过通常的 java.net 接口。

- Android.os：封装了系统服务、消息传输、IPC 机制。
- Android.opengl：封装了 opengl 的工具、3D 加速。
- Android.provider：封装了类访问 Android 的内容提供者。
- Android.telephony：封装了与拨打电话相关的 API 交互。
- Android.view：封装了基础的用户界面接口框架。
- Android.util：涉及工具性的方法，例如时间日期的操作。
- Android.webkit：默认浏览器操作接口。
- Android.widget：封装了各种 UI 元素（大部分是可见的），在应用程序的屏幕中使用。

2. Android 在线帮助文档

Android 的官方网站提供了 Android 在线帮助文档，这是用户进行程序设计的好工具，希望大家都能用好这个工具。在 Android 的官方网站上提供了目前最新的在线帮助文档，其网址为 http://developer.android.com/reference/packages.html，如图 1.11 所示。

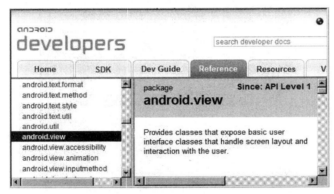

图 1.11　Android 在线帮助文档

1.4　Android 应用程序的开发过程

1.4.1　开发 Android 应用程序的一般过程

开发 Android 应用程序的一般过程如图 1.12 所示。

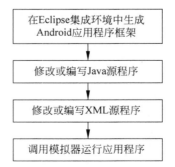

视频演示

图 1.12　Android 应用程序的开发过程

1.4.2 生成 Android 应用程序框架

1. 创建一个新的 Android 项目

启动 Eclipse，选择"文件"（File）→"新建"（New）→"项目"（Project）命令，如果已经安装了 Android 的 Eclipse 插件，将会在弹出的如图 1.13 所示的"新建项目"对话框中看到 Android 的选项。选择 Android Application Project，单击"下一步"按钮。

2. 填写应用程序的参数

在 New Android App 对话框中输入应用程序名称、项目名称、包名等参数，并选择 Android SDK 的版本，如图 1.14 所示。

图 1.13　新建一个 Android 项目

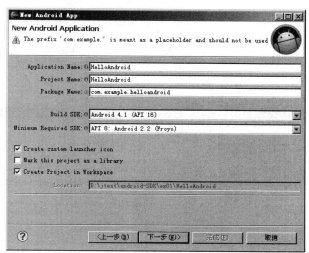

图 1.14　输入项目参数并选择 Android SDK 的版本

该对话框中参数的含义如下。

- Application Name：项目的应用程序名称。
- Project Name：项目名称，系统默认为与应用程序相同。
- Package Name：包名，遵循 Java 规范，用包名来区分不同的类是很重要的，示例中所用的包名是 com.example.helloandroid。

3. 填写相关程序参数

一个 Android 项目至少由两个程序组成：一个是项目的主程序（即首先执行的程序），其扩展名为.java；另一个是显示用户界面的程序，其扩展名为.xml。这里需要填写这两个程序的相关数据，如图 1.15 所示。

该对话框中参数的含义如下。

- Activity Name：项目的主程序（项目的入口程序）名称，其扩展名为.java。
- Layout Name：界面布局程序的名称，其扩展名为.xml。
- Title：显示的标题。

单击图 1.15 所示的对话框中的"完成"按钮，系统会自动生成一个 Android 应用项目框架，如图 1.16 所示。

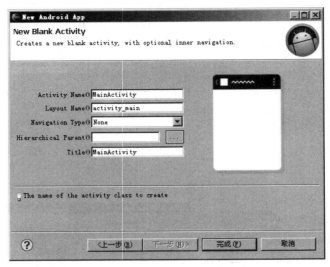

图 1.15 输入显示页面参数

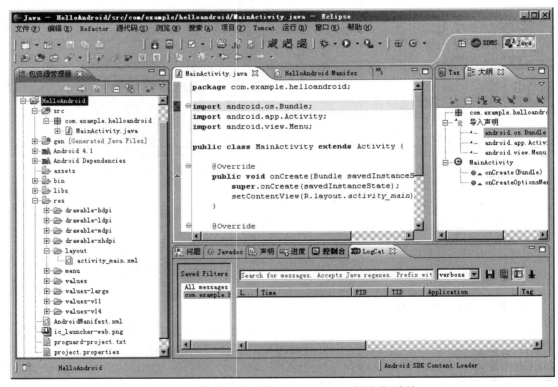

图 1.16 系统自动生成的 HelloAndroid 应用项目框架

1.4.3 编写 MainActivity.java

创建 HelloAndroid 项目后，打开主程序 MainActivity.java 文件，可以看到系统自动生成了以下代码：

```
1  package com.example.helloandroid;
2  import android.app.Activity;
3  import android.os.Bundle;
4  public class MainActivity extends Activity      ← 主程序是 Activity 的子类
5  { /** Called when the activity is first created. */
6    @Override
7    public void onCreate(Bundle savedInstanceState)
8    {
9        super.onCreate(savedInstanceState);
10       setContentView(R.layout.activity_main);    ← 显示 activity_main.xml 定
11   }                                                  义的用户界面
12 }
```

在 Android 系统中，应用程序的入口程序（主程序）都是活动程序界面 Activity 类的子类。在上述代码中，最重要的是第 10 行，其显示了 activity_main.xml 定义的用户界面。

1.4.4 配置应用程序的运行参数

（1）在包资源管理器中右击项目名称 HelloAndroid，在弹出的快捷菜单中选择"运行方式"→"运行 配置"命令，如图 1.17 所示。

（2）在弹出的"运行 配置"对话框中选择 Android 选项卡，单击 Browse 按钮，然后选择需要运行的 HelloAndroid 项目，如图 1.18 所示。

图 1.17 选择"运行 配置"命令

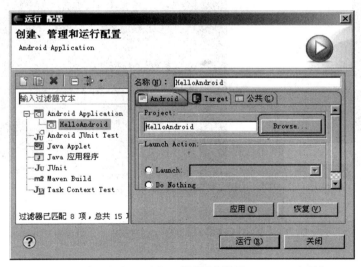

图 1.18 在"运行 配置"对话框中设置 Android 选项卡

（3）在"运行 配置"对话框中选择 Target 选项卡，然后选择事先已经设置的模拟器 AVD 设备，如图 1.19 所示。

图 1.19　在 Target 选项卡中选择模拟器 AVD 设备

1.4.5　在模拟器中运行应用程序

单击工具栏中的"运行 Android Application"按钮 ▶，运行 AVD 模拟器，可以看到应用程序的运行结果（首次运行程序时可能耗时较长），如图 1.20 所示。

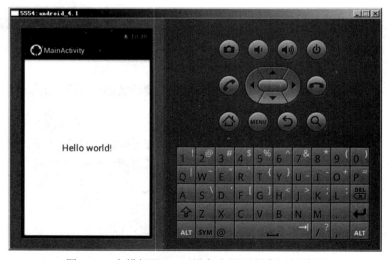

图 1.20　在模拟器 AVD 设备上显示程序运行结果

1.5　Android 应用程序结构

1.5.1　目录结构

打开 HelloAndroid 项目，在包资源管理器中可以看到应用项目的目录和文件结构，如图 1.21 所示。

实际上，图 1.21 所示的目录结构内容是最基本的，程序员还可以在此基础上添加需要的内容。下面对 Android 项目结构的基本内容进行介绍。

1. src 目录

src 目录存放 Android 应用程序的 Java 源代码文件。在系统自动生成的项目结构中，有一个在创建项目时输入 Create Activity 名称的 Java 文件，即 MainActivity.java，其源代码如图 1.22 所示。

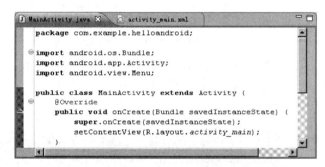

图 1.21　HelloAndroid 项目的目录和文件结构

图 1.22　src 目录下的 MainActivity.java 的源代码

2. res 目录及资源类型

Android 系统的资源为应用项目所需要的声音、图片、视频、用户界面文档等，其资源文件存放于项目的 res 目录下。res 的目录结构及资源类型如表 1-2 所示。

表 1-2　Android 系统的资源目录结构及类型

目录结构	资源类型
res/values	存放字符串、颜色、尺寸、数组、主题、类型等资源
res/layout	XML 布局文件
res/drawable	图片（BMP、PNG、GIF、JPG 等）
res/anim	XML 格式的动画资源（帧动画和补间动画）
res/menu	菜单资源
res/raw	可以放任意类型的文件，一般存放比较大的音频、视频、图片或文档，会在 R 类中生成资源 ID，封装在 APK 中
assets	可以存放任意类型的文件，不会被编译，与 RAW 相比，不会在 R 类中生成资源 ID

（1）drawable 细分为 drawable-hdpi、drawable-ldpi、drawable-mdpi、drawable-xhdpi 子目录，分别存放分辨率大小不同的图标资源，以便相同的应用程序在分辨率大小不同的显示窗体上都可以顺利显示。系统开始运行时，会检测显示窗体的分辨率大小，自动选择与

显示窗体分辨率大小匹配的目录，获取大小匹配的图标，如表 1-3 所示。

表 1-3 4 种分辨率大小不同的图标

子 目 录	图标分辨率大小	图 例
drawable-xhdpi	96×96	
drawable-hdpi	72×72	
drawable-mdpi	48×48	
drawable-ldpi	36×36	

（2）在 layout 子目录中存放用户界面布局文件。其中有一个系统自动生成的 activity_main.xml 文件，它可以按可视化的图形设计界面显示，也可以按代码设计界面显示，如图 1.23（a）和（b）所示。

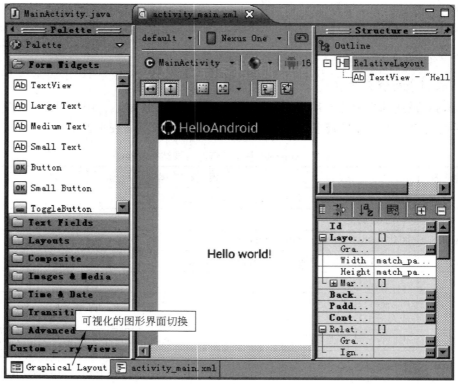

（a）图形设计界面

```
┌─────────────────────────────────────────────────────────┐
│ MainActivity.java │ activity_main.xml  ✕               │
├─────────────────────────────────────────────────────────┤
│ ⊟ <RelativeLayout xmlns:android="http://schemas.androi  │
│       xmlns:tools="http://schemas.android.com/tools"    │
│       android:layout_width="match_parent"               │
│       android:layout_height="match_parent" >            │
│       <TextView                                         │
│           android:layout_width="wrap_content"           │
│           android:layout_height="wrap_content"          │
│           android:layout_centerHorizontal="true"        │
│           android:layout_centerVertical="true"          │
│           android:padding="@dimen/padding_medium"       │
│           android:text="@string/hello_world"            │
│           android:textSize="24sp"                       │
│           tools:context=".MainActivity" />              │
│   </RelativeLayout>                                     │
├─────────────────────────────────────────────────────────┤
│ Graphical Layout │ activity_main.xml                    │
└─────────────────────────────────────────────────────────┘
```

(b) 代码设计界面

图 1.23　用户界面布局文件 activity_main.xml

activity_main.xml 布局文件的代码如下：

```
1  <?xml version="1.0" encoding="utf-8"?>
2  <LinearLayout xmlns:android="http://schemas.android.com/apk/res/android"
3      android:orientation="vertical"
4      android:layout_width="fill_parent"
5      android:layout_height="fill_parent"
6      >
7  <TextView
8      android:layout_width="fill_parent"
9      android:layout_height="wrap_content"
10     android:text="@string/hello"
11     />
12 </LinearLayout>
```

对布局文件中的参数解析如下。

- <LinearLayout>：线性布局配置，在这个标签中，所有元件都是按由上到下的顺序排列的。
- android: orientation：表示这个介质的布局配置方式是从上到下垂直地排列其内部视图。
- android: layout_width：定义当前视图在屏幕上所占的宽度，fill_parent 表示填充整个屏幕。
- android: layout_height：定义当前视图在屏幕上所占的高度。

在应用程序中需要使用用户界面的组件时，需要通过 gen 目录下的 R.java 文件中的 R 类来调用。

（3）values 子目录存放参数描述文件资源。这些参数描述文件也都是 XML 文件，如字符串（string.xml）、颜色（color.xml）、数组（arrays.xml）等。这些参数同样需要通过 gen 目录下的 R.java 文件中的 R 类来调用。

3. gen 目录

gen 目录存放由 ADT 系统自动产生的 R.java 文件，该文件将 res 目录中的资源与 ID 编号进行映射，从而可以方便地对资源进行引用，如图 1.24 所示。正如该文件头部的说明，该文件是自动生成、不允许用户修改的。

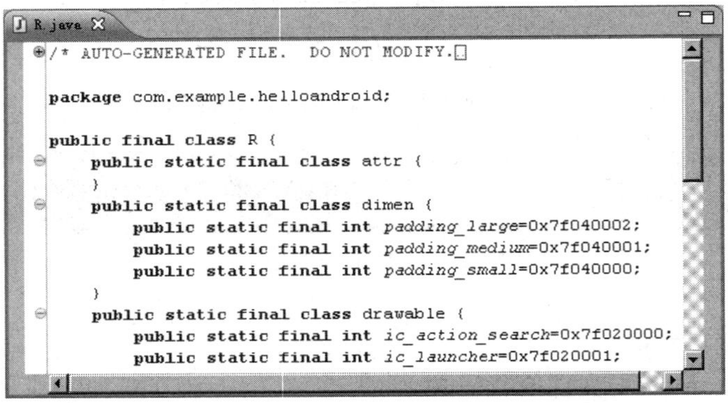

图 1.24　gen 目录下的 R.java 的源代码

在程序中引用资源需要使用 R 类，其引用格式如下：

`R.资源文件类型.资源名称`

例如：

（1）在 activity 中显示布局视图：

`setContentView(R.layout.activity_main);`

（2）程序中的组件对象 mButtn 与用户界面布局文件中的按钮对象 Button1 建立关联：

`mButtn=(Button)finadViewById(R.id.Button1);`

（3）程序中的组件对象 mEditText 与用户界面布局文件中的组件对象 EditText1 建立关联：

`mEditText=(EditText)findViewById(R.id.EditText1);`

程序中的 findViewById(int id)方法是 Java 控制程序中的组件对象与用户界面程序组件对象进行关联的"桥梁"，如图 1.25 所示。

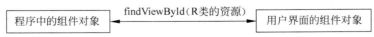

图 1.25　Java 程序中的组件对象与用户界面的组件对象进行关联

4. AndroidManifest.xml 项目配置文件

AndroidManifest.xml 是每个应用程序都需要的系统配置文件，它位于应用程序根目录下。AndroidManifest.xml 相当于一个注册表文件，Android 应用程序的应用组件及使用的权限都必须在这个文件中声明。例如，在配置文件中声明使用网络的权限、使用摄像头的权限等，以便系统运行应用程序时能正常调用这些组件或设备。

系统自动生成的 AndroidManifest.xml 文件的代码如下：

```
1   <?xml version="1.0" encoding="utf-8"?>
2   <manifest xmlns:android="http://schemas.android.com/apk/res/android"
3       package="com.HelloAndroid"
4       android:versionCode="1"
5       android:versionName="1.0">
6       <uses-sdk android:minSdkVersion="14"/>
7       <application
8           android:icon="@drawable/ic_launcher"
9           android:label="@string/app_name">
10          <activity
11              android:label="@string/app_name"
12              android:name=".MainAndroidActivity">
13              <intent-filter>
14                  <action android:name="android.intent.action.MAIN"/>
15                  <category android:name="android.intent.category.LAUNCHER"/>
16              </intent-filter>
17          </activity>
18      </application>
19  </manifest>
```

AndroidManifest.xml 文件中代码元素的意义见表 1-4。

表 1-4 AndroidManifest.xml 文件代码说明

代 码 元 素	说　　明
manifest	XML 文件的根节点，包含了 package 中所有的内容
xmlns:android	命名空间的声明 xmlns:android="http://schemas.android.com/apk/res/android" 使得 Android 中的各种标准属性能在文件中使用
package	声明应用程序包
uses-sdk	声明应用程序所使用的 Android SDK 版本
application	application 级别组件的根节点，声明一些全局或默认的属性，如标签、图标、必要的权限等
android:icon	应用程序图标
android:label	应用程序名称
activity	Activity 是一个应用程序与用户交互的图形界面，每一个 Activity 必须有一个 <activity>标记对应，如果一个 Activity 没有对应的标记，将无法运行
android:name	应用程序默认启动的活动程序 Activity 界面

续表

代码元素	说 明
intent-filter	声明一组组件支持的 Intent 值，在 Android 中，组件之间可以相互调用、协调工作，Intent 提供组件之间通信所需要的相关信息
action	声明目标组件执行的 Intent 动作，Android 定义了一系列标准动作，如 MAIN_ACTION、VIEW_ACTION、EDIT_ACTION 等，与此 Intent 匹配的 Activity 将会被当作进入应用的入口
category	指定目标组件支持的 Intent 类别

AndroidManifest.xml 文件的一般结构如图 1.26 所示，其中，声明使用的权限以及 service 组件在后面的章节会有详细介绍。

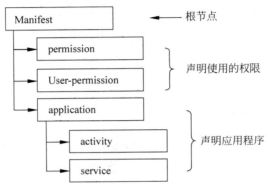

图 1.26　AndroidManifest.xml 文件的一般结构

1.5.2　Android 应用程序架构分析

1. 逻辑控制层与表现层

从上面的 Android 应用程序可以看到，一个 Android 应用程序通常由 Activity 程序（Java 程序）和用户界面布局文件组成。

在 Android 应用程序中，逻辑控制层与表现层是分开设计的。逻辑控制层由 Java 程序实现，表现层由 XML 文档描述，如图 1.27 所示。

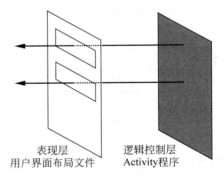

图 1.27　Android 应用程序的逻辑控制层与表现层

2. Android 程序的组成结构

Android 程序与 Java 程序的结构是相同的，打开 src 目录下的 HelloAndroid.java 文件，其代码如下：

```
1  package com.example.HelloAndroid;        ← 包声明语句
2  import android.app.Activity;              ← 导入包
3  import android.os.Bundle;
```

```
 4   public class MainAndroid extends Activity          ← 类标志
                                                         ← 类声明语句
 5   {                                                   ← 类名
 6       public void onCreate(Bundle savedInstanceState) ← 重写 onCreate()方法
 7       {
 8           super.onCreate(savedInstanceState);         ← 调用父类 Activity 的 onCreate()方法
 9           setContentView(R.layout. activity_main);
10       }                                               ← 在屏幕上显示内容的方法
11   }
```

语句说明：

（1）第 1 行是包声明语句，包名是在建立应用程序时指定的。在这里设定为：

`package com.example.HelloAndroid;`

这一行的作用是指出这个文档所在的名称空间。package（包）是其关键字。使用名称空间的原因是程序一旦扩展到某个大小，程序中的变量名称、方法名称、类名等难免重复，这时就可以通过定义名称空间，将定义的名称区隔，以避免相互冲突的情形发生。

（2）第 2、3 行是导入包的声明语句。这两条语句的作用是告诉系统编译器，编译程序时要导入 android.app.Activity 和 android.os.Bundle 两个包。import（导入）是其关键字。在 Java 语言中，使用任何 API 都要事先导入相对应的包。

（3）第 4～11 行是类的定义，这是应用程序的主体部分。Android 应用程序是由类组成的，类的一般结构为：

```
public class MainAndroid extends Activity    //类声明
{
    … ;                                       //类体
}
```

class 是类的关键字，HelloAndroid 是类名。在 public class MainAndroid 后面添加 extends Activity，表示 MainAndroid 类继承 Activity 类。这时，称 Activity 类是 HelloAndroid 类的父类，或称 HelloAndroid 类是 Activity 类的子类。extends 是表示继承关系的关键字。在面向对象的程序中，子类会继承父类的所有方法和属性。也就是说，对于在父类中所定义的全部方法和属性，子类可以直接使用。由于 Activity 类是一个具有屏幕显示功能的活动界面程序，因此，其子类 MainAndroid 也具有屏幕显示功能。

class 语句后面的一对大括号"{ }"表示复合语句，它是该类的主体部分，称为类体。在类体中定义类的方法和变量。

（4）第 6～10 行是在 MainAndroid 类的类体中定义一个方法。

1.6 Android 应用程序设计示例

【例 1-1】在模拟器中显示"我对学习 Android 很感兴趣！"。

（1）在 Eclipse 中新建一个 Android 项目，其 Application Name（项目名称）为 Ex01_01、

Package Name（包名）为 com.ex01_01。

（2）在系统自动生成的应用程序框架中，打开修改资源目录 res\values 中的字符串文件 string.xml，其代码如下：

```
1  <resources>
2    <string name="app_name">HelloAndroid</string>
3    <string name="hello_world">Hello world!</string>      ← 修改该行代码
4    <string name="menu_settings">Settings</string>
5    <string name="title_activity_main">MainActivity</string>
6  </resources>
```

在第 3 行代码中找到 XML 文档元素：

```
<string name="hello_world">Hello world!</string>
```

将其修改为：

```
<string name="hello_world">我对学习 Android 很感兴趣!</string>
```

（3）保存程序。选择"运行"→"运行配置"命令，运行项目。在模拟器中的运行结果如图 1.28 所示。

图 1.28　程序在模拟器中运行的显示结果

【例 1-2】设计一个显示资源目录中图片文件的程序。

（1）在 Eclipse 中新建一个 Android 项目，其 Application Name（项目名称）为 Ex01_02、Package Name（包名）为 com.ex01_02。

（2）把事先准备的图片文件 flower.png 复制到资源目录 res\drawable-hdpi 中，如图 1.29（a）所示。

视频演示

（3）打开源代码目录 src 中的 MainActivity.java 文件，编写代码如下：

```
1  package com.ex01_02;
2  import android.app.Activity;
3  import android.os.Bundle;
4  import android.widget.ImageView;      ← 增加导入 ImageView 类的语句
5  public class MainActivity extends Activity {
6    /** Called when the activity is first created. */
7    @Override
8    public void onCreate(Bundle savedInstanceState) {
9      super.onCreate(savedInstanceState);
10     //setContentView(R.layout.activity_main);      ← 注释该语句
```

```
11    ImageView img=new ImageView(this);         ← 创建 ImageView 对象并实例化
12    img.setImageResource(R.drawable.flower);   ← ImageView 对象设置
13    setContentView(img);   ← 把 ImageView 对象     引用图片资源
14  }                         显示到屏幕上
15 }
```

（4）保存程序。选择"运行"→"运行配置"命令，运行项目。在模拟器中的运行结果如图1.29（b）所示。

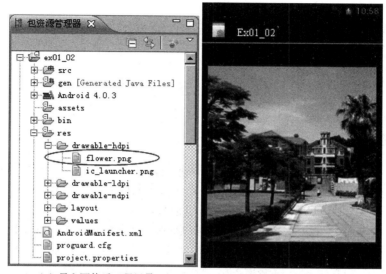

(a) 导入图片后工程目录　　(b) 程序在模拟器中运行的结果

图 1.29　在模拟器中显示图片

习　题　1

1. Android 系统是基于什么操作系统的应用系统？
2. 试述建立 Android 系统开发环境的过程和步骤。
3. 如何编写和运行 Android 系统应用程序？
4. 编写 Android 应用程序，在模拟器中显示"我对 Android 很痴迷！"。
5. 编写 Android 应用程序，在模拟器中显示一个图形文件。

第 2 章　Java 语法概述

由于 Android 系统采用 Java 语言编写应用程序，所以本章对 Android 系统中所使用的 Java 语言基础知识进行简明扼要的介绍。

2.1　语 法 基 础

2.1.1　数据类型

程序在执行过程中，需要对数据进行运算，还需要存储数据。这些数据可能是由使用者输入的，也可能是从文件中取得的，甚至是从网络上得到的。在程序运行的过程中，这些数据通过变量（Variable）存储在内存中，以便程序随时调用。

数据存储在内存的一块空间中，为了取得数据，必须知道这块内存空间的位置，然而若使用内存地址编号，则相当不方便，所以通常用一个变量名来表示。变量（Variable）是一个数据存储空间的表示，将数据指定给变量，就是将数据存储至对应的内存空间，调用变量，就是将对应的内存空间的数据取出来使用。

一个变量代表一个内存空间，数据就存储在这个空间中，然而由于数据在存储时所需要的容量各不相同，不同的数据需要分配不同大小的内存空间来存储。在 Java 语言中，对于不同的数据用不同的数据类型（Data Type）来区分。

Android 系统所使用的数据类型可以分为两大类：基本数据类型和引用数据类型。基本数据类型是由程序设计语言系统所定义、不可再划分的数据类型。基本数据类型的数据所占内存的大小固定，与软/硬件环境无关。基本数据类型在内存中存入的是数据值本身。引用数据类型在内存中存入的是指向该数据的地址，不是数据本身，它往往由多个基本数据组成。因此，对引用数据类型的应用称为对象引用，引用数据类型也称为复合数据类型，在有些程序设计语言中称为指针。

Android 系统有 8 个基本数据类型：字节型（byte）、短整型（short）、整型（int）、长整型（long）、字符型（char）、单精度型（float）、双精度型（double）、布尔型（boolean），这些类型可分为 4 组。

- 整数型：该组包括字节型（byte）、短整型（short）、整型（int）、长整型（long），它们有符号整数。
- 实数型：该组包括单精度型（float）、双精度型（double），它们代表有小数精度要求的数字。实数型又称为浮点型。
- 字符型：该组包括字符型（char），它代表字符集的符号，如字母和数字。

- 布尔型：该组包括布尔型（boolean），它是一种特殊的类型，表示真/假值。

下面是 Android 系统常用数据类型的分类，如图 2.1 所示。

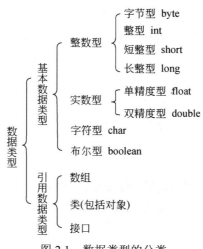

图 2.1　数据类型的分类

每一种具体数据类型都对应着唯一的类型关键字、类型长度和值域范围，见表 2-1。

表 2-1　Android 系统的基本数据类型

类　型	数据类型关键字	适　用　于	类型长度	值　域　范　围
字节型	byte	非常小的整数	1	$-128 \sim 127$
短整型	short	较小的整数	2	$-2^{15} \sim 2^{15}-1$ 内的整数
整型	int	一般整数	4	$-2^{31} \sim 2^{31}-1$ 内的整数
长整型	long	非常大的整数	4	$-2^{31} \sim 2^{31}-1$ 内的整数
单精度型	float	一般实数	4	$-3.402823 \times 10^{38} \sim 3.402823 \times 10^{38}$ 内的数
双精度型	double	非常大的实数	8	$-1.7977 \times 10^{308} \sim 1.7977 \times 10^{308}$ 内的数
字符型	char	单个字符	1	
布尔型	boolean	判断	1	true 和 false

2.1.2　常量与变量

在程序中，每一个数据都有一个名字，并且在内存中占据一定的存储单元。在程序运行过程中，数据值不能改变的量称为常量，其值可以改变的量称为变量。

在 Android 系统中，所有常量及变量在使用前必须先声明其值的数据类型，也就是要遵守"先声明后使用"的原则。声明变量的作用有两点：一是确定该变量的标识符（变量名），以便系统为其指定存储地址和识别它，这便是"按名访问"原则；二是为该变量指定数据类型，以便系统为其分配足够的存储单元。

声明变量的格式为：

    数据类型　变量名1, 变量名2, …　；

例如：

```
int a;                //a 的值在程序运行过程中可能发生变化,将其声明为变量
int x, y, sum;        //同时声明多个变量,变量之间用逗号分隔
```

在 Android 系统中，常量的声明与变量的声明非常类似，例如：

```
final int DAY=24;           //DAY 的值在整个程序中保持不变,将其声明为常量
final double PI=3.14159;    //声明圆周率常数
```

从上面的示例中可以看出，常量声明的前面多了一个关键字 final，并且赋了一个固定值。在习惯上，常量用大写字母，变量用小写字母，以示区别。

2.1.3 对变量赋值

在程序中经常需要对变量赋值，在 Android 程序中用赋值号（＝）表示。所谓赋值，就是把赋值号右边的数据或运算结果赋给左边的变量。其一般格式为：

> 变量 = 表达式;

例如：

```
int x=5;          //指定 x 为整型变量,并赋初值 5
char c='a';       //指定 a 为字符型变量,并赋初值 'a'
```

如果同时对多个相同类型的变量赋值，可以用逗号分隔，例如：

```
int x=5, y=8, sum;
sum=x+y;          //将 x+y 的运算结果赋给变量 sum
```

在 Android 程序中，经常会用到形如"x = x + a"的赋值运算，例如：

```
int x=5;
x=x+2;
```

其中，右边 x 的值是 5，加 2 后，把运算结果 7 赋给左边的变量 x，所以 x 的值是 7。
注意："x = x + a"常常简写成"x += a"。

2.1.4 关键字

所谓关键字就是 Android 系统中已经规定了特定意义的单词。它们用来表示一种数据类型，或者表示程序的结构等。注意，不可以把这些单词用作常量名或变量名。

Android 系统中规定的关键字有 abstract、boolean、break、byte、case、catch、char、class、continue、default、do、double、else、extends、false、final、finally、float、for、if、implements、import、instanceof、int、interface、long、native、new、null、package、private、protected、public、return、short、static、super、switch、synchronized、this、throw、throws、transient、true、try、void、volatile、while。

2.1.5 转义符

在 Android 系统中提供了一些特殊的字符常量，这些特殊字符称为转义符。通过转义

符可以在字符串中插入一些无法直接输入的字符，如换行符、引号等。每个转义符都以反斜杠（\）为标志。例如，"\n"代表一个换行符，这里的"n"不再代表字母 n，而是"换行"符号。常用的以"\"开头的转义符见表 2-2。

表 2-2 常用转义符

转 义 符	含 义
\b	退格
\f	走纸换页
\n	换行
\r	回车
\t	横向跳格（Ctrl+I）
\'	单引号
\"	双引号
\\	反斜杠

2.2 基本数据类型应用示例

2.2.1 整型与浮点型

1. 整型

当用变量表示整数时，通常将变量声明为整型。

【例 2-1】用整型变量计算两个数的和。

视频演示

```
1   /*  计算两个数的和  */
2   package com.ex02_01;
3
4   import android.app.Activity;
5   import android.os.Bundle;
6   import android.widget.TextView;
7   public class Ex02_01Activity extends Activity {
8
9     public void onCreate(Bundle savedInstanceState) {
10
11      super.onCreate(savedInstanceState);
12      //setContentView(R.layout.activity_main);
13      int x, y, sum;          ← 声明 3 个整型变量
14      x=3;                    
15      y=5;                    ← 给变量 x、y 赋值
16      sum=x+y;                ← 求和
17      TextView txt=new TextView(this);
18      txt.setText( "x=3;"+"\n y=5;"+"\n x+y="+sum);
19      setContentView(txt);    ← 在屏幕上显示结果
20    }
21  }
```

程序的运行结果如图 2.2 所示。

语句说明：

在程序的第 13 行声明了 3 个整型变量：x、y、sum，这 3 个变量与存储器中相应类型的存储单元对应。在程序运行到第 14 行语句时，数值 3 存放到被编译器命名为 x 的内存单元中；运行到第 15 行语句时，数值 5 存放到被编译器命名为 y 的内存单元中；运行到第 16 行语句时，将内存单元 x 和内存单元 y 中的值相加并将结果放到变量 sum 中，如图 2.3 所示。

图 2.2 计算两个数的和

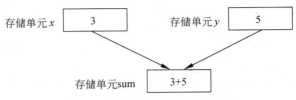

图 2.3 进行加法运算的内存单元

2. 浮点型

【例 2-2】用双精度浮点型变量计算一个圆的面积。

```
1   /*计算圆的面积*/
2   package com.ex02_02;
3
4   import android.app.Activity;
5   import android.os.Bundle;
6   import android.widget.TextView;
7   public class Ex02_02Activity extends Activity {
8
9       public void onCreate(Bundle savedInstanceState) {
10
11          super.onCreate(savedInstanceState);
12          //setContentView(R.layout.activity_main);
13          double pi, r, s;          ← 声明 3 个双精度浮点型变量
14          r=10.8;                   ← 给变量赋值
15          pi=3.14159;
16          s=pi * r * r;             ← 计算圆的面积
17          TextView txt=new TextView(this);
18          txt.setText( "圆的面积为： " + s);
19          setContentView(txt);      ← 在屏幕上显示结果
20      }
21  }
```

程序的运行结果如图 2.4 所示。

图 2.4 计算圆的面积

2.2.2 字符型

1. 字符变量

在 Android 系统中，存储字符的数据类型是 char，一个字符变量每次只能存放一个字符。一个字符变量在内存中占用两个字节（16 位）的存储空间，也就是说，一个字符变量可以存放两个字节长度的字符。例如，一个汉字是两字节长度，一个字符变量可以存放一个汉字。但当存放的字符是 8 位时，例如，一个字母是一个字节长度（8 位），一个字符变量只能存放一个字母。

【例 2-3】演示字符型变量的用法。

```
1   /*字符型变量的用法*/
2   package com.ex02_03;
3
4   import android.app.Activity;
5   import android.os.Bundle;
6   import android.widget.TextView;
7   public class Ex02_03Activity extends Activity {
8
9       public void onCreate(Bundle savedInstanceState) {
10
11          super.onCreate(savedInstanceState);
12          //setContentView(R.layout.activity_main);
13          char ch1,ch2,ch3;              //声明3个字符型变量
14          ch1=88; //88 在 Unicode 码中对应的是字母 X;
15          ch2='Y';                        //给变量赋值
16          ch3='汉';
17          TextView txt=new TextView(this);
18          txt.setText( " ch1、ch2、ch3: " + ch1 + "、" + ch2+ "、" +ch3);
19          setContentView(txt);           //在屏幕上显示结果
20      }
21  }
```

程序的运行结果如图 2.5 所示。

2. 字符串

用双引号括起来的多个（也可以是一个或空）字符常量称为字符串。字符串是程序设计中最常用的数据类型。

例如：

"我对 Android 很痴迷！\n" 和 "a+b=" 等都是字符串。

字符串与字符有以下区别：

字符是由单引号括起来的单个字符，而字符串是由双引号括起来的，且可以是零个或多个字符。例如，'abc' 是不合法的，"" 是合法的，表示空字符串。

在 Android 系统中，用 String 来定义字符串，String 是 Android 系统的一个类，其使用

图 2.5 字符型变量的用法

方法见例 2-4。

在 Android 系统中，还经常用 CharSequence 定义字符串的类型，其用法基本与 String 相同。

【例 2-4】 字符串的用法示例。

```
1    /*字符串的用法*/
2    package com.ex02_04;
3
4    import android.app.Activity;
5    import android.os.Bundle;
6    import android.widget.TextView;
7    public class Ex02_04Activity extends Activity {
8
9        public void onCreate(Bundle savedInstanceState) {
10
11           super.onCreate(savedInstanceState);
12           //setContentView(R.layout.activity_main);
13           String str;                    ◄── 声明字符串对象
14           str="This is a String!";        ◄── 给字符串变量赋值
15           TextView txt=new TextView(this);
16           txt.setText(str);
17           setContentView(txt);            ◄── 在屏幕上显示结果
18       }
19   }
```

程序的运行结果如图 2.6 所示。

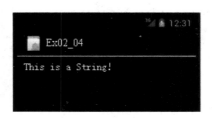

图 2.6　字符串的用法

2.2.3　布尔型

Java 中表示逻辑值的基本类型称为布尔型，它只有 true 和 false 两个值，且它们不对应于任何整数值。例如：

```
boolean b=true;
```

布尔型是所有诸如 $a<b$ 这样的关系运算的返回类型，它对管理控制语句的条件表达式也是必需的。

【例 2-5】 布尔型变量的用法示例。

```
1   /*布尔型变量的用法*/
2   package com.ex02_05;
3
4   import android.app.Activity;
5   import android.os.Bundle;
6   import android.widget.TextView;
7   public class Ex02_05Activity extends Activity {
8
9     public void onCreate(Bundle savedInstanceState) {
10
11      super.onCreate(savedInstanceState);
12      //setContentView(R.layout.activity_main);
13      boolean a, b, c;          ← 声明 3 个布尔型变量
14      a=true;
15      b=false;                  ← 给变量赋值
16      c=10<8;
17      TextView txt=new TextView(this);
18      txt.setText("a is " + a + "\n b is " + b + "\n10 < 8 is " + c);
19      setContentView(txt);      ← 在屏幕上显示结果
20    }
21  }
```

程序的运行结果如图 2.7 所示。

图 2.7　布尔型变量的用法

2.2.4　数据类型的转换

在 Java 语言中，对于已经定义了类型的变量，允许转换其类型。变量的数据类型转换分为自动类型转换和强制类型转换两种。

1. 自动类型转换

在程序中已经对变量定义了一种数据类型，若想以另一种数据类型表示时，要符合以下两个条件：

（1）转换前的数据类型与转换后的数据类型兼容。
（2）转换后的数据类型比转换前的数据类型表示的范围大。
对于基本数据类型按精度从"低"到"高"的顺序为：

```
byte→short→int→long→float→double
低                              高
```

当把级别低的变量的值赋给级别高的值时,系统自动进行数据类型转换。
例如:

```
int  x=10;
float y;
y=x;
```

这时 y 的值为 10.0。

2. 强制类型转换

强制类型转换是指把级别高的变量的值赋给级别低的变量。其转换格式为:

(类型名) 要转换的值或变量;

例如:设有

```
int a;
double b=3.14;
```
则
```
a=(int)b;
```
◀── 将 b 强制类型转换为 int 类型后,再赋值给 a

结果 $a = 3$,b 仍然是 double 类型,b 的值仍然是 3.14。

从该示例可以看出,采用强制类型转换将高类型数据转换成低类型数据时,可能会降低数据的精确度。

2.3 程序控制语句

2.3.1 语句的分类

语句组成了一个执行程序的基本单元,它类似于自然语言的句子。Java 语言的语句可以分为以下几类:

1. 表达式语句

```
x=3;
y=5;
sum=x+y;
```

在一个表达式的最后加上一个分号就构成了一个语句,分号是语句不可缺少的部分。

2. 复合语句

用{ }把一些语句括起来就构成了复合语句。

```
{
    String str="我对 Android 很痴迷!";
    TextView txt=new TextView(this);
    txt.setText(str);
}
```

3. 控制语句

控制语句用于控制程序流程及执行的先后顺序，主要有顺序控制语句、条件控制语句和循环控制语句 3 种类型。

4. 包语句和引用语句

包语句和引用语句是 Android 系统提供的一种类名管理机制。在介绍类和接口之后，将详细讲解包语句和引用语句的用法。

2.3.2 顺序控制语句

顺序控制是指计算机在执行这种结构的程序时，从第一条语句开始，按从上到下的顺序依次执行程序中的每一条语句。顺序控制是程序最基本的结构，包含选择控制语句和循环控制语句的程序，在总体执行上也是按顺序结构执行的。

【例 2-6】交换两个变量的值。

在编写程序时，有时需要把两个变量的值互换，交换值的运算需要用到一个中间变量。例如，要将 *a* 与 *b* 的值互换，可用下面这样一段程序：

其中，temp 是中间变量，它仅起过渡作用。交换过程如图 2.8 所示。

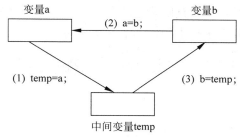

图 2.8　a、b 两数的交换

程序代码如下：

```
1    /*交换两个变量的值*/
2    package com.ex02_06;
3
4    import android.app.Activity;
5    import android.os.Bundle;
6    import android.widget.TextView;
7    public class Ex02_06Activity extends Activity {
8
9        public void onCreate(Bundle savedInstanceState) {
10
```

```
11        super.onCreate(savedInstanceState);
12        //setContentView(R.layout.activity_main);
13        int a=3, b=5, temp;         ← 声明3个整型变量
14        temp=a;
15        a=b;                         ← 交换变量a、b的值
16        b=temp;
17        TextView txt=new TextView(this);
18        txt.setText("a="+a+"\t  b="+b);
19        setContentView(txt);         ← 在屏幕上显示结果
20    }
21 }
```

程序的运行结果如图2.9所示。

图2.9 交换两个变量的值

2.3.3 if 语句

1. 单分支选择结构

if 语句用于实现选择结构。它判断给定的条件是否满足，并根据判断结果决定执行某个分支的程序段。对于单分支选择语句，其语法格式为：

这个语法的意思是，当条件表达式给定的条件成立时(true)，执行其中的语句块，若条件不成立(false)，则跳过这部分语句，直接执行后续语句。

其流程如图2.10所示。

【例2-7】设有任意3个数 a、b、c，按从小到大的顺序依次输出这3个数。

首先将 a 与 b 比较，如果 a<b，本身就是从小到大排列的，则保持原顺序不变。但如果 a>b，则需要交换 a、b 两个变量的值。依此类推，再将 a 与 c 比较、b 与 c 比较，最后得到从小到大排序的结果。其算法流程如图2.11所示。

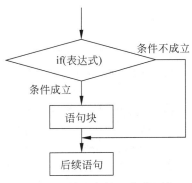

图2.10 单分支的 if 条件语句

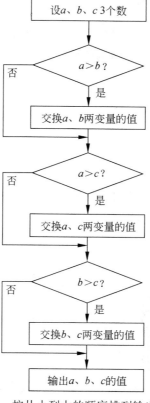

图 2.11 按从小到大的顺序排列输出两数

程序代码如下：

```
1   /*对任意3个整数，按从小到大的顺序排列*/
2   package com.ex02_07;
3
4   import android.app.Activity;
5   import android.os.Bundle;
6   import android.widget.TextView;
7   public class Ex02_07Activity extends Activity {
8
9       public void onCreate(Bundle savedInstanceState) {
10
11          super.onCreate(savedInstanceState);
12          // setContentView(R.layout.activity_main);
13          int a=9, b=5, c=7, t;                      ◄── 声明3个整型变量
14          if(a>b){ t=a;  a=b;   b=t;}
15          if(a>c){ t=a;  a=c;   c=t;}                ◄── 排序
16          if(b>c){ t=b;  b=c;   c=t;}
17          TextView txt = new TextView(this);
18          txt.setText("a="+a+"\t  b="+b+"\t c="+c);
19          setContentView(txt);                       ◄── 在屏幕上显示结果
20      }
21  }
```

程序的运行结果如图 2.12 所示。
2. 双分支选择结构

有时需要在条件表达式不成立的时候执行不同的语句，可以使用双分支选择结构的条件语句，即 if-else 语句。双分支选择结构的语法格式为：

```
if(表达式)
   { 语句块 1;}
else
   { 语句块 2;}
```

该语法的意思是，当条件表达式成立时（true），执行语句块 1，否则（else）执行语句块 2。对于双分支选择类型的条件语句，其流程如图 2.13 所示。

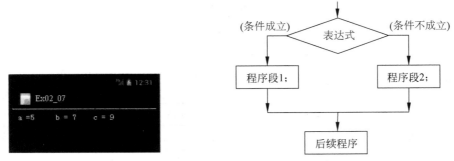

图 2.12　对任意 3 个整数，按从小到大的顺序排列　　图 2.13　双分支选择结构的条件语句

if-else 语句的扩充格式是 if-else if。一个 if 语句可以有任意个 if-else if 部分，但只能有一个 else。

2.3.4　switch 语句

switch 语句是一个多分支选择语句，也称开关语句。它可以根据一个整型表达式有条件地选择一个语句执行。if 语句只有两个分支可以选择，而实际问题中常常需要用到多分支的选择，当然也可以用嵌套 if 语句来处理，但如果分支较多，则嵌套的 if 语句层数太多，会造成程序冗长且执行效率降低。

switch 的语法结构形式如下：

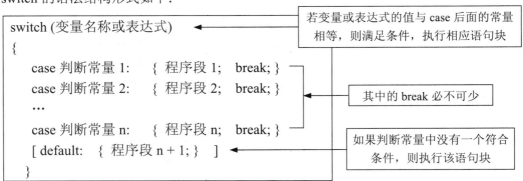

switch 语句首先计算条件表达式的值，如果表达式的值和某个 case 后面的判断常量相同，则执行该 case 里的若干条语句，直到 break 语句为止。若没有一个判断常量相同，则执行 default 后面的若干条语句。default 语句块可以没有。在 case 语句块中，break 是必不可少的，break 表示终止 switch，跳转到 switch 的后续语句继续运行。

switch 语句的流程如图 2.14 所示。

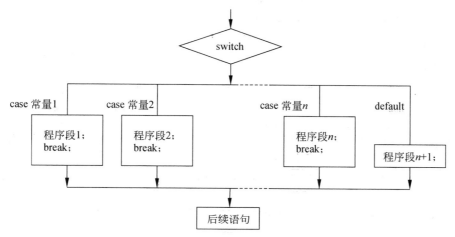

图 2.14 switch 语句的流程

2.3.5 循环语句

在程序设计过程中，经常需要将一些功能按一定的要求重复执行多次，我们将这一过程称为循环。

循环结构是程序设计中一种很重要的结构。其特点是，在给定条件成立时，反复执行某程序段，直到条件不成立为止。给定的条件称为循环条件，反复执行的程序段称为循环体。

1. for 循环语句

for 循环语句的语法结构如下：

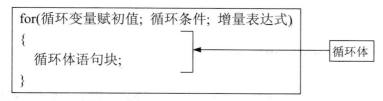

在 for 语句中，其语法成分是：

（1）循环变量赋初值是初始循环的表达式，它在循环开始的时候就被执行一次。

（2）循环条件决定什么时候终止循环，这个表达式在每次循环的过程中被计算一次。当表达式的计算结果为 false 的时候，这个循环结束。

（3）增量表达式是每循环一次循环变量增加多少（即步长）的表达式。

（4）循环体是被重复执行的程序段。

for 语句的执行过程是这样的：首先执行循环变量赋初值，完成必要的初始化工作；再判断循环条件，若循环条件满足，则进入循环体中执行循环体的语句；执行完循环体后，接着执行 for 语句中的增量表达式，以便改变循环条件，这一轮循环就结束了。第二轮循环又从判断循环条件开始，若循环条件仍能满足，则继续循环，否则跳出整个 for 语句，执行后续语句，如图 2.15 所示。

图 2.15 循环语句的执行过程

【例 2-8】求从 1 加到 100 的和。

```
1   /*利用循环求和*/
2   package com.ex02_08;
3
4   import android.app.Activity;
5   import android.os.Bundle;
6   import android.widget.TextView;
7   public class Ex02_08Activity extends Activity {
8
9     public void onCreate(Bundle savedInstanceState) {
10
11      super.onCreate(savedInstanceState);
12      //setContentView(R.layout.activity_main);
13      int sum=0;         ← 变量 sum 存放累加值，初始值为 0
14      for(int i=1; i<=100; i++)   ← i 为循环变量，每循环一次，i 自加 1（步长为 i++），循环终止条件为 i>100
15      {
16        sum=sum+i;       ← 循环体内，每循环一次，累加一次循环变量的值
17      }
18      TextView txt=new TextView(this);
19      txt.setText("1+2+3+...+100="+sum);
20      setContentView(txt);   ← 在屏幕上显示结果
21    }
22  }
```

语句说明：

在程序中，i 是改变条件表达式的循环变量。在开始循环之初，循环变量 $i=1$、sum=0，这时 $i<100$，满足循环条件，因此可以进入循环体，执行第 10 行累加语句：sum+i=1+0=1，将结果再放回到变量 sum 中，完成第一次循环。接着，循环变量自加 1（i++），此时，$i=2$，再和循环条件比较，……，如此反复，sum=sum+i 一直累加，直到运行了 100 次，$i=101$，循环条件 $i<=100$ 不再满足，循环结束。

程序的运行结果如图 2.16 所示。

图 2.16 利用循环求和

2. while 循环语句

Android 系统有两种 while 循环语句，即 while 语句和 do-while 语句。这两种循环结构

的流程如图 2.17 所示。

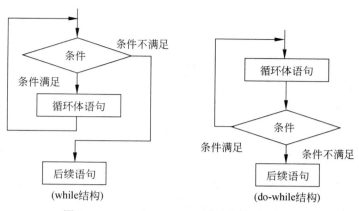

图 2.17 while 和 do-while 循环结构的流程图

1) while 语句

while 语句的基本语法结构为：

```
while(循环条件表达式)
{
   … 循环体;
}
```

首先，while 语句执行条件表达式，返回一个 boolean 值（true 或者 false）。如果条件表达式返回 true，则执行大括号中的循环体语句。然后继续测试条件表达式并执行循环体代码，直到条件表达式返回 false。

2) do-while 语句

do-while 语句的基本语法结构为：

```
do
 {
   …循环体;
 } while(循环条件表达式);
```

do-while 语句与 while 语句的区别在于，语句先执行循环中的语句再计算条件表达式，所以 do-while 语句的循环体至少被执行一次。

【例 2-9】计算 1! + 2! + 3! + … + 10!

算法分析：这是一个多项式求和问题。每一项都是计算阶乘，可以利用循环结构来处理。
其代码如下：

```
1   /*while 循环*/
2   package com.ex02_09;
3
4   import android.app.Activity;
```

```
5    import android.os.Bundle;
6    import android.widget.TextView;
7    public class Ex02_09Activity extends Activity {
8
9      public void onCreate(Bundle savedInstanceState) {
10
11       super.onCreate(savedInstanceState);
12       //setContentView(R.layout.activity_main);
13       int sum=0, i=1, p=1;
14       do
15       {
16          p=p*i;         ← 计算阶乘         ┐
17          sum=sum+p;     ← 累加              ├ do-while 结构的循环体
18          i++;           ← 循环变量自增      │
19       } while(i<=10);   ← 循环条件         ┘
20       TextView txt=new TextView(this);
21       txt.setText("1!+2!+3!+...+100!="+sum);
22       setContentView(txt);
23     }
24   }
```

程序的运行结果如图 2.18 所示。

图 2.18 while 循环应用

3. 循环嵌套

循环可以嵌套，在一个循环体内包含另一个完整的循环，称为循环嵌套。循环嵌套运行时，外循环每执行一次，内层循环都要执行一个周期。

【例 2-10】应用循环嵌套，编写一个按 9 行 9 列排列输出的乘法九九表程序。

算法分析：用双重循环控制乘法九九表按 9 行 9 列排列输出，用外循环变量 i 控制行数，i 从 1 到 9 取值。内循环变量 j 控制列数，由于 $i*j=j*i$，故内循环变量 j 没有必要从 1 到 9 取值，只需从 1 到 i 取值就够了。外循环变量 i 每执行一次，内循环变量 j 执行 i 次。

其代码如下：

```
1   /*循环嵌套应用*/
2   package com.ex02_10;
3
4   import android.app.Activity;
```

```
5   import android.os.Bundle;
6   import android.widget.TextView;
7   public class Ex02_10Activity extends Activity {
8
9     public void onCreate(Bundle savedInstanceState) {
10
11        super.onCreate(savedInstanceState);
12        TextView txt=new TextView(this);
13      int i,j;
14      for(i=1; i<=9;i++)
15      {
16       for(j=1; j<=i;j++)
17       {
18          txt.append(i+"x"+j+"="+i*j+"\t");
19       }
20       txt.append("\n");
21      }
22      setContentView(txt);
23    }
24  }
```

程序的运行结果如图 2.19 所示。

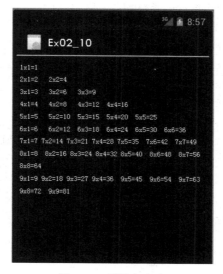

图 2.19 乘法九九表

1×1=1
2×1=2   2×2=4
3×1=3   3×2=6   3×3=9
4×1=4   4×2=8   4×3=12   4×4=16
5×1=5   5×2=10   5×3=15   5×4=20   5×5=25

```
6×1=6    6×2=12   6×3=18   6×4=24   6×5=30   6×6=36
7×1=7    7×2=14   7×3=21   7×4=28   7×5=35   7×6=42   7×7=49
8×1=8    8×2=16   8×3=24   8×4=32   8×5=40   8×6=48   8×7=56   8×8=64
9×1=9    9×2=18   9×3=27   9×4=36   9×5=45   9×6=54   9×7=63   9×8=72   9×9=81
```

循环可以嵌套，可以是　　　　　，但不能交叉，即　　　　　是不允许的。

2.3.6 跳转语句

Android 程序设计中主要应用了两种跳转语句：break 语句、continue 语句。

1. break 语句

break 语句有两种作用：其一，break 语句被用来退出 switch 结构，跳出 switch 结构继续执行后续语句；其二，break 语句被用来中止循环。

在循环体中使用 break 语句强行退出循环时，忽略循环体中的任何其他语句和循环的条件测试，终止整个循环，程序跳到循环后面的语句继续运行。

【例 2-11】使用 break 语句跳出循环。

其代码如下：

```
1   /*使用break语句跳出循环*/
2   package com.ex02_11;
3   import android.app.Activity;
4   import android.os.Bundle;
5   import android.widget.TextView;
6   public class Ex02_11Activity extends Activity
7   {
8     public void onCreate(Bundle savedInstanceState)
9     {
10      super.onCreate(savedInstanceState);
11      //setContentView(R.layout.activity_main);
12      TextView txt=new TextView(this);
13      String output="";
14      for(int i=0; i<100; i++)
15        {
16          if(i==10) {break;}          ← i=10 时跳出循环
17          output=output+"i="+i+"\n";
18        }
19      output=output+"\n 循环 10 次后，跳出循环！";
20      txt.setText(output);
21      setContentView(txt);
```

```
22    }
23 }
```

语句说明：

循环变量 i 的取值从 0 开始，当 i=10 时，满足第 10 行 if 语句的条件，运行 break 语句，跳出循环，转向执行第 13 行的语句（注意，最后一次执行循环体时第 11 行的语句没被执行）。

程序的运行结果如图 2.20 所示。

2. continue 语句

continue 语句用来终止本次循环。其功能是终止当前正在进行的本轮循环，即跳过后面剩余的语句，转而执行循环的第一条语句，计算和判断循环条件，决定是否进入下一轮循环。

【例 2-12】应用 continue 语句输出用 "*" 排列的三角形。

其代码如下：

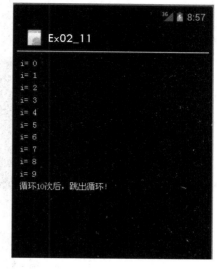

图 2.20 使用 break 语句跳出循环

```
1  /*应用continue语句输出用"*"排列的三角形*/
2  package com.ex02_12;
3  import android.app.Activity;
4  import android.os.Bundle;
5  import android.widget.TextView;
6  public class Ex02_12Activity extends Activity
7  {
8    public void onCreate(Bundle savedInstanceState)
9    {
10     super.onCreate(savedInstanceState);
11     //setContentView(R.layout.activity_main);
12     TextView txt=new TextView(this);
13     String output="";
14     for(int i=0; i<5; i++) {
15       for(int j=0; j<5; j++) {
16         if(j>i) {
17           continue;          ← 终止内循环    内循环控制    外循环控
18         }                                    行*号个数    制行数
19         output=output+ "*"+" ";
20       }
21       output=output+"\n";
22     }
23     txt.append(output);
24     setContentView(txt);
25   }
26 }
```

语句说明：

在本例中第 17 行的 continue 语句终止了内循环，而跳转到第 6 行继续执行外循环的下一轮循环。

程序的运行结果如图 2.21 所示。

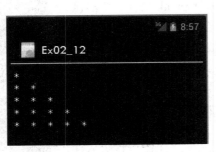

图 2.21 输出用"*"排列的三角形

2.4 类 与 对 象

类和对象是 Android 的核心和本质。它们是 Java 语言的基础，编写一个 Java 程序，在某种程度上来说就是定义类和创建对象。定义类和建立对象是 Java 编程的主要任务。

2.4.1 类的定义

从本书开始我们就接触到类了。类是组成 Java 程序的基本要素，在本节将介绍如何创建一个类。

1. 类的一般形式

类由类声明和类体组成，而类体又由成员变量和成员方法组成，即：

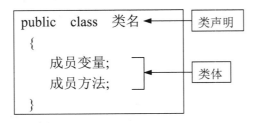

2. 类的继承性

继承性是面向对象的程序中两个类之间的一种关系，即一个类可以从另一个类（即它的父类）继承状态和行为。被继承的类（父类）也可以称为超类，继承父类的类称为子类。继承为组织和构造程序提供了一个强大而自然的机理。

面向对象系统允许一个类建立在其他类之上。例如，山地自行车、赛车及双人自行车都是自行车，那么在面向对象技术中，山地自行车、赛车及双人自行车就是自行车类的子类，自行车类是山地自行车、赛车及双人自行车的父类。

子类声明的一般形式如下：

```
class 类名 [extends 父类名] [implements 接口列表]
{
    …
}
```

对各组成部分的具体说明如下：

1）类的关键字 class

在类声明中，class 是声明类的关键字，表示类声明的开始，类声明后面跟着类名，按习惯类名要用大写字母开头，并且类名不能用阿拉伯数字开头。给类命名时，最好取一个容易识别且有意义的名字，避免 A、B、C 之类的类名。

2）声明父类

extends 为声明该类的父类，表明该类是其父类的子类。一个子类可以从它的父类继承变量和方法。

创建子类的格式如下：

```
class SubClass extends 父类名
{
    …
}
```

3）实现接口

为了在类声明中实现接口，要使用关键字 implements，并且在其后面给出接口名。当要实现有多个接口时，各接口名以逗号分隔，其形式为：

```
implements 接口 1, 接口 2, …
```

接口是一种特殊的抽象类，这种抽象类中只包含常量和方法的定义，而没有变量和方法的实现。一个类可以实现多个接口，以某种程度实现"多继承"。

【例 2-13】创建一个 Activity 的子类。

其代码如下：

```
1   /*类声明示例*/
2   package com.ex02_13;
3
4   import android.app.Activity;
5   import android.os.Bundle;
6   import android.widget.TextView;
7   public class Ex02_13Activity extends Activity        ← 声明继承父类 Activity
8   {
9       int x, y, sum;                                    ← 声明成员变量
10      public void onCreate(Bundle savedInstanceState)   ← 定义成员方法
11      {
12          super.onCreate(savedInstanceState);           ← 调用父类 Activity 的方法
13
```

```
14        x=3;              ← 给变量 x、y 赋值
15        y=5;
16        sum=x+y;  ← 求和
17        TextView txt=new TextView(this);
18        txt.setText(" x=3;" + "\n y=5;"+"\n x+y="+sum);
19        setContentView(txt);   ← 在屏幕上显示结果
20    }
21 }
```

2.4.2 对象

大家知道，类是一个抽象的概念，而对象是类的具体化。类与对象的关系相当于普通数据类型与其变量的关系。声明一个类只是定义了一种新的数据类型，类通过实例化创建了对象，才真正创建了这种数据类型的物理实体。

1. 对象的创建

创建对象的一般格式为：

> 类名 对象名 = new 类名([参数列表]);

该表达式隐含了 3 个部分：对象的声明、实例化和初始化。

1）对象的声明

声明对象的一般形式为：

> 类名 对象名;

声明对象并不为对象分配内存空间，而只是分配一个引用空间；对象的引用类似于指针，是 32 位的地址空间，它的值指向一个中间的数据结构，其存储有关数据类型的信息以及当前对象所在堆的地址，而对于对象所在的实际的内存地址是不可操作的，这就保证了安全性。

2）实例化

实例化是为对象分配内存空间和进行初始化的过程，其一般形式为：

> 对象名 = new 构造方法( );

运算符 new 为对象分配内存空间，它调用对象的构造方法，返回引用；一个类的不同对象分别占据不同的内存空间。在执行类的构造方法进行初始化时，可以根据参数类型或个数调用相应的构造方法，进行不同的初始化，实现方法重构。

2. 对象的使用

在前面介绍了类，那么如何使用类呢？类是不能直接使用的，我们使用的是类通过实例化成为的对象。而对象是通过访问对象变量或调用对象方法来使用的。

通过运算符"."可以实现对对象的变量访问和方法的调用。变量和方法可以通过设定访问权限来限制其他对象对它的访问。

1）访问对象的变量

对象创建之后，对象就有了自己的变量。对象通过使用运算符"."实现对自己的变量

访问。

访问对象成员变量的格式为：

对象名.成员变量;

例如，设有一个 A 类，其结构如下：

```
class A
  { int x; }
```

如果要对其变量 *x* 赋值，则先创建并实例化类 A 的对象 a，然后再通过对象给变量 *x*：

```
A a=new A();
  a.x=5;
```

2）调用对象的方法

对象通过使用运算符 "." 实现对自己的方法调用。

调用对象成员方法的格式为：

对象名.方法名([参数列表]);

例如在例 2-14 中，定义了 Box 类。在 Box 类中定义了 3 个 double 类型的成员变量和一个 volume()方法，将来每个具体对象的内存空间中都保存有自己的 3 个变量和一个方法的引用，并由它的 volume()方法来操纵自己的变量，这就是面向对象的封装特性的体现。要访问或调用一个对象的变量或方法需要首先创建这个对象，然后用算符 "." 调用该对象的某个变量或方法。

【**例 2-14**】应用创建类的实例对象计算长方体的体积。

其代码如下：

```
1    /*构造长方体*/
2    package com.ex02_14;
3    import android.app.Activity;
4    import android.os.Bundle;
5    import android.widget.TextView;
6    public class Ex02_14Activity extends Activity
7    {
8      public void onCreate(Bundle savedInstanceState)
9      {
10       super.onCreate(savedInstanceState);
11       Box box=new Box();          ◄── 应用构造方法创建实例对象
12       TextView txt=new TextView(this);   ◄── 实例化对象
13       double v;
14       v=box.volume();             ◄── 调用对象的普通方法
15       txt.setText("长方体体积为: " + v);
16       txt.setTextSize(25);
17       setContentView(txt);
18     }
```

```
19
20  /*创建一个内部类*/
21  class Box
22  {
23      double width, height, depth;
24      Box()
25        {
26            width=10;
27            height=10;
28            depth=10;
29        }
30      double volume()
31        {
32            return width * height * depth;
33        }
34  }
35  }
```

第 26~28 行 → Box 类的构造方法，与类同名

第 32 行 → 用普通方法计算长方体体积

语句说明：

（1）在本例中第 21~34 行创建了一个内部类 Box，其中，第 24~29 行为构造方法。构造方法的特点是方法名与类名相同，且在类的前面无返回类型。第 30~33 行为普通方法。

（2）程序的第 11 行应用"对象名 = new 构造方法()"创建 Box 类的实例对象 box。

（3）程序的第 12 行创建 TextView 类的实例对象 txt。

（4）程序的第 14 行为调用对象的普通方法。

程序的运行结果如图 2.22 所示。

图 2.22　创建实例化对象的示例

2.4.3　接口

接口是类的一种（抽象类），只包含常量和方法的定义，没有变量和具体方法的实现，且其方法都是抽象方法。它的用处体现在以下几个方面：

（1）通过接口实现不相关类的相同行为，而无须考虑这些类之间的关系。

（2）通过接口指明多个类需要实现的方法。

（3）通过接口了解对象的交互界面，而无须了解对象所对应的类。

1. 接口定义的一般格式

接口的定义包括接口声明和接口体。

接口定义的一般格式如下：

```
[public] interface 接口名 [extends 父接口名]
{
    …    //接口体
}
```

extends 子句与类声明的 extends 子句基本相同，不同的是，一个接口可有多个父接口，用逗号隔开，而一个类只能有一个父类。

2. 接口的实现

在类的声明中用 implements 子句来表示一个类使用某个接口，在类体中可以使用接口中定义的常量，而且必须实现接口中定义的所有方法。一个类可以实现多个接口，在 implements 子句中用逗号隔开。

2.4.4 包

在 Java 语言中，每个类都会生成一个字节码文件，该字节码文件名与类名相同。这样，可能会发生同名类的冲突。为了解决这个问题，Java 采用包来管理类名空间。包不仅提供了一种类名管理机制，还提供了一种面向对象方法的封装机制。包将类和接口封装在一起，方便了类和接口的管理与调用。例如：Java 的基础类都封装在 java.lang 包中，所有与网络相关的类都封装在 java.net 包中，等等。程序设计人员也可以将自己编写的类和接口根据需要封装到一个包中。

1. 包的定义

把一个源程序归入到某个包的方法用 package 来实现。

package 语句的一般格式为：

```
package 包名;
```

例如，要编写一个 MyTest.java 源文件，并且文件存放在当前运行目录的子目录 abc\test 下，则：

```
package abc.test;
public class MyTest
{
    ...
}
```

在源文件中，package 是源程序的第一条语句。包名一定是当前运行目录的子目录。一个包内的 Java 代码可以访问该包的所有类及类中的非私有变量和方法。

2. 包的引用

如果要使用包中的类，必须用关键字 import 导入这些类所在的包。

import 语句的一般格式为：

```
import 包名.类名;
```

当要引用包中所有的类或接口时，类名可以用通配符"*"代替。

2.5　XML 语法简介

XML（Extensible Markup Language，可扩展标记语言）是一套定义语义标记的规则，这些标记将文档分成许多部分并对这些部分加以标识。XML 的语法规则既简单又严格，熟悉 HTML 的读者会发现它的语法和 HTML 很相似，非常容易学习和使用。

1. XML 文档结构

下面来看一个 XML 文档实例，其代码如下：

```
1  <?xml version="1.0" encoding=" utf-8"?>
2  <bookstore>
3      <book category="计算机">
4          <title lang="中文">Java语言程序设计</title>
5          <author>张思民</author>
6          <year>2012</year>
7          <price>39.00</price>
8      </book>
9  </bookstore>
```

语句说明：

（1）第 1 行是 XML 声明，描述文档定义的版本是 XML1.0 版和所使用的编码方式是 utf-8 编码。

（2）第 2 行定义文档的根元素为< bookstore>，根元素类似 HTML 文档中的<HTML>开头标记。

（3）第 3 行定义根的子元素<book>，并定义了 book 的属性 category="计算机"。

（4）第 4~7 行分别定义元素<book>的 4 个子元素（title、author、year 及 price）。

（5）第 8 行定义元素的结尾</book>。

（6）第 9 行定义根元素的结尾</bookstore>。

从以上实例可以看出，XML 文档由文档声明、元素、属性、文本、实体、注释等内容组成。

XML 文档是一种树结构，XML 文档必须包含一个根元素。从"根部"开始，然后扩展到"枝叶"部分。上例描述一本书的 XML 文档结构如图 2.23 所示。

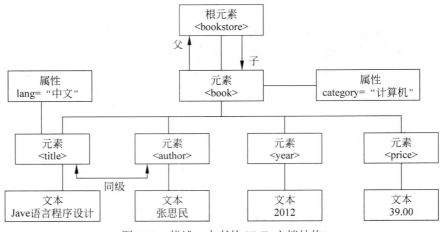

图 2.23　描述一本书的 XML 文档结构

2. 元素

元素是 XML 文档的基本组成部分，元素其实就是标记内容。XML 文档中共有 4 类元素：空元素、仅含文本的元素、包含其他元素的元素、混合元素。

（1）空元素。如果一个元素中没有任何文本内容，那么它就是一个空元素。例如：

`<book> </book>`

（2）仅含文本的元素。有些元素中仅含文本内容，例如：

`<author>张思民</author>`

（3）包含其他元素的元素。一个元素中可以包含其他元素。该元素称为父元素，被包含的元素称为子元素。例如：

```
<book category="计算机">
    <title lang="中文">Java语言程序设计</title>
    <author>张思民</author>
    <year>2012</year>
    <price>39.00</price>
</book>
```

（4）混合元素。混合元素既包含文本内容又包含子元素。

3. 属性

XML 元素可以拥有属性。属性是对标识进行进一步的描述和说明，一个标识可以有多个属性。在 XML 中，属性值必须用单引号或双引号括起来，其基本格式为：

`<元素名　属性名 = "属性值">`

例如：

`<title lang="中文">`

4. 注释

注释以"<!--"开始，以"-->"结束，注释内的任何标记都会被忽略。注释可以出现在 XML 文档的任何位置。其基本格式为：

`<!-- 注释内容 -->`

【例 2-15】系统自动生成的应用程序界面布局文件 main.xml 语句分析。

main.xml 布局文件的代码如下：

```
1  <?xml version="1.0" encoding="utf-8"?>
2  <LinearLayout xmlns:android="http://schemas.android.com/apk/res/android"
3      android:orientation="vertical"
4      android:layout_width="fill_parent"
5      android:layout_height="fill_parent"
6      >
7  <TextView
8      android:layout_width="fill_parent"
9      android:layout_height="wrap_content"
10     android:text="@string/hello"
11     />
12 </LinearLayout>
```

语句说明：

（1）第 1 行定义 XML 文档声明，说明该文档的版本是 XML1.0 版，所使用的编码方式是 utf-8 编码。

（2）第 2 行定义根元素<LinearLayout>，到第 12 行结束，该元素说明界面布局的排列方式。其中，xmlns 为根元素的属性，其属性是一个名称为 android 的命名空间，其值是固定的：

xmlns:android="http://schemas.android.com/apk/res/android"

该网址中有该文件所使用的全部元素的定义。在编写该文件时，如果不注明命名空间，编译器并不会报错，但在程序运行时可能会发生错误。

（3）第 3～5 行均为根元素的属性值。第 3 行说明布局按从上到下的垂直方式排列组件；第 4 行定义布局宽度；第 5 行定义布局高度。

（4）第 7～11 行定义元素<TextView>，该元素是一个文本组件，第 8～10 行均为说明该元素的属性。第 8、9 行定义文本组件的宽和高，第 10 行定义文本组件的文本内容。

【例 2-16】系统自动生成的应用程序配置文件 AndroidManifest.xml 语句分析。

AndroidManifest.xml 文件的代码如下：

```
1   <?xml version="1.0" encoding="utf-8"?>
2   <manifest xmlns:android="http://schemas.android.com/apk/res/android"
3     package="com.HelloAndroid"
4     android:versionCode="1"
5     android:versionName="1.0" >
6     <uses-sdk android:minSdkVersion="15" />
7     <application
8       android:icon="@drawable/ic_launcher"
9       android:label="@string/app_name" >
10      <activity
11        android:label="@string/app_name"
12        android:name=".HelloAndroidActivity" >
13        <intent-filter >
14          <action android:name="android.intent.action.MAIN" />
15          <category android:name="android.intent.category.LAUNCHER"/>
16        </intent-filter>
17      </activity>
18    </application>
19  </manifest>
```

语句说明：

（1）第 1 行定义 XML 文档声明，说明该文档的版本是 XML1.0 版，所使用的编码方式是 utf-8 编码。

（2）第 2 行定义根元素<manifest>，到第 19 行结束。xmlns:android 为命名空间属性。

（3）第 3、4、5 行定义根元素属性。第 3 行指定应用程序唯一的包名 package，该应用程序的包名为"com.HelloAndroid"。第 12 行 activity 元素指定应用程序名称为.HelloAndroidActivity，这只是简化名称，完整的应用程序名称应该加上包名，即 com.HelloAndroid.HelloAndroidActivity。

（4）第 6 行指定运行的最低版本号。

（5）第 7～18 行定义子元素<application>。

（6）第 10～17 行定义 application 的子元素<activity>。

（7）第 13～16 行定义 activity 的子元素<intent-filter>，该元素为指定应用程序的启动条件和运行程序的入口。

习 题 2

1. 设有 3 个数：$a = 5$，$b = 8$，$c = 3$，按从大到小的顺序排列显示。
2. 编写显示下列图形的程序：

```
     （1）              （2）              （3）
      #             * * * * * * *           $
      ##             * * * * *              $ $
      ###             * * *                $ $ $ $ $
      ####             *                    $ $ $
                                             $
```

3. 分析下列程序，写出运行结果。

```java
package com.ex02_test;
import android.app.Activity;
import android.os.Bundle;
import android.widget.TextView;

public class MainActivity extends Activity
{
    TextView txt;
    ShowMessage sm;                  //声明接口变量
    @Override
    public void onCreate(Bundle savedInstanceState)
    {
        super.onCreate(savedInstanceState);
        txt=new TextView(this);
        sm=new TV();                 //接口变量中存放对象的引用
        sm.显示商标("长城牌电视机");   //接口回调
        setContentView(txt);
    }

    interface  ShowMessage
    {
        void 显示商标(String s);
    }
```

```
class TV implements ShowMessage
{   public void 显示商标(String s)
    {
        txt.setText(s);
    }
}
```

第 3 章　Android 用户界面设计

3.1　用户界面组件包 widget 和 View 类

1. 用户界面组件包 widget

Android 系统为开发人员提供了丰富多彩的用户界面组件，通过使用这些组件可以设计出炫丽的界面。大多数用户界面组件放置在 android.widget 包中。widget 包中的常用组件如表 3-1 所示。

表 3-1　widget 包中的常用组件

可视化组件	说　　明
Button	按钮
CalendarView	日历视图
CheckBox	复选框
EditText	文本编辑框
ImageView	显示图像或图标，并提供缩放、着色等各种图像处理方法
ListView	列表框视图
MapView	地图视图
RadioGroup	单选按钮组
Spinner	下拉列表
TextView	文本标签
WebView	网页浏览器视图
Toast	消息提示

2. View 类

View 是用户界面组件的共同父类，几乎所有的用户界面组件都是继承 View 类实现的，如 TextView、Button、EditText 等。

对于 View 类及其子类的属性，可以在界面布局文件中设置，也可以通过成员方法在 Java 代码文件中动态设置。View 类的常用属性和方法如表 3-2 所示。

表 3-2　View 类的常用属性和方法

属　　性	对 应 方 法	说　　明
android:background	setBackgroundColor (int color)	设置背景颜色
android:id	setId(int)	为组件设置可通过 findViewById 方法获取的标识符

续表

属　性	对 应 方 法	说　　明
android:alpha	setAlpha(float)	设置透明度，取值范围为 0～1
	findViewById(int id)	与 id 所对应的组件建立关联
android:visibility	setVisibility(int)	设置组件的可见性
android:clickable	setClickable(boolean)	设置组件是否响应单击事件

3.2　文本标签与按钮

3.2.1　文本标签

文本标签（TextView）用于显示文本内容，是最常用的组件之一。其常用方法见表 3-3。

表 3-3　文本标签（TextView）常用方法

方　　法	功　　能
getText();	获取文本标签的文本内容
setText(CharSequence text);	设置文本标签的文本内容
setTextSize(float);	设置文本标签的文本大小
setTextColor(int color);	设置文本标签的文本颜色

其常用的 XML 文件元素属性见表 3-4。

表 3-4　文本标签（TextView）常用的 XML 文件元素属性

元　素　属　性	说　　明
android:id	文本标签标识
android:layout_width	文本标签（TextView）的宽度，通常取值"fill_parent"（屏幕宽度）或以像素为单位 pt 的固定值
android:layout_height	文本标签（TextView）的高度，通常取值"wrap_content"（文本的高）或以像素 px 为单位的固定值
android:text	文本标签（TextView）的文本内容
android:textSize	文本标签（TextView）的文本大小

【例 3-1】设计一个文本标签组件程序。

创建名为 Ex03_01 的新项目，包名为 com.ex03_01。打开系统自动生成的项目框架，需要设计的文件为：

- 界面布局文件 activity_main.xml；
- 控制文件 MainActivity.java；
- 资源文件 strings.xml。

视频演示

（1）设计界面布局文件 activity_main.xml。在界面布局文件 activity_main.xml 中加入文本标签 TextView，设置文本标签组件的 id 属性，如图 3.1 所示。

activity_main.xml 代码如下：

```
1  <?xml version="1.0" encoding="utf-8"?>
2  <LinearLayout xmlns:android="http://schemas.android.com/apk/res/android"
3      android:layout_width="fill_parent"
4      android:layout_height="fill_parent"
5      android:orientation="vertical" >
6      <TextView
7          android:id="@+id/textView1"            ← 设置文本标签的 id 属性值
8          android:layout_width="fill_parent"
9          android:layout_height="wrap_content"
10         android:text="@string/hello" />
11 </LinearLayout>
```

图 3.1　在界面布局中设置文本标签

（2）设计控制文件 MainActivity.java。在控制文件 MainActivity.java 中添加文本标签组件，并将界面布局文件中所定义的文本标签元素属性值赋给文本标签，与界面布局文件中的文本标签建立关联。程序代码如下：

```
1  package com.ex03_01;
2  import android.app.Activity;
3  import android.os.Bundle;
4  import android.graphics.Color;      ← 引用图形颜色组件
5  import android.widget.TextView;     ← 引用文本标签组件
6
7  public class MainActivity extends Activity
```

Android 用户界面设计

```
 8  {
 9    private TextView  txt;          ◀── 声明文本标签对象
10    public void onCreate(Bundle savedInstanceState)
11    {
12      super.onCreate(savedInstanceState);
13      setContentView(R.layout.activity_main);
14      txt=(TextView)findViewById(R.id.textView1);   ◀── 与界面布局文件中的文本标签建立关联
15      txt.setTextColor(Color.WHITE);   ◀── 设置文本颜色
16    }
17  }
```

（3）设计资源文件 strings.xml。修改资源文件 strings.xml 中属性为"hello"的元素项的文本内容：

```
 1  <?xml version="1.0" encoding="utf-8"?>
 2  <resources>
 3      <string name="hello">\n     荷塘月色
 4                                  \n 剪一段时光缓缓流淌，
 5                                  \n 流进了月色中微微荡漾，
 6                                  \n 弹一首小荷淡淡的香，
 7                                  \n 美丽的琴音就落在我身旁.
 8      </string>
 9      <string name="app_name">Ex03_01</string>
10  </resources>
```

保存项目，配置应用程序的运行参数。程序的运行结果如图 3.2 所示。

图 3.2　文本标签

3.2.2　按钮

按钮（Button）用于处理人机交互事件，在一般应用程序中经常会用到。由于按钮是文本标签（TextView）的子类，其继承关系如图 3.3 所示。按钮继承了文本标签的所有方法和属性。

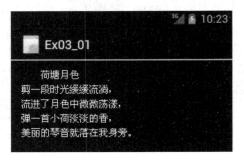

图 3.3　按钮与文本标签的继承关系

按钮在程序设计中最常用的方式是实现 OnClickListener 监听接口，当单击按钮时，通过 OnClickListener 监听接口触发 onClick()事件，实现用户需要的功能。OnClickListener 接口有一个 onClick()方法,在按钮实现 OnClickListener 接口时,一定要重写这个方法。

按钮调用 OnClickListener 接口对象的方法如下:

按钮对象.setOnClickListener(OnClickListener 对象);

【例3-2】编写程序,实现单击按钮页面标题及文本标签的文字内容发生变化的功能,如图 3.4 所示。

　　单击按钮前　　　　单击按钮后　　　　　视频演示

图 3.4　单击按钮后,文本标签的文字内容发生变化

创建名为 Ex03_02 的新项目,包名为 com.ex03_02。

(1)设计界面布局文件 activity_main.xml。在界面布局文件中添加一个按钮,将其 id 设置为 button1。其代码如下:

```
1  <?xml version="1.0" encoding="utf-8"?>
2  <LinearLayout xmlns:android="http://schemas.android.com/apk/res/android"
3      android:layout_width="fill_parent"
4      android:layout_height="fill_parent"
5      android:orientation="vertical" >
6      <TextView
7          android:id="@+id/textView1"           ← 设置文本标签的 id 属性值
8          android:layout_width="fill_parent"
9          android:layout_height="wrap_content"
10         android:text="@string/hello" />
11     <Button
12         android:id="@+id/button1"             ← 设置按钮的 id 属性值
13         android:layout_width="fill_parent"
14         android:layout_height="wrap_content"
15         android:text=" @string/button" />
16 </LinearLayout>
```

(2)设计控制文件 MainActivity.java。在控制文件 MainActivity.java 中设计一个实现按钮监听接口的内部类 mClick,当单击按钮时,触发 onClick()事件。其代码如下:

```
1  package com.ex03_02;
2  import android.app.Activity;
3  import android.os.Bundle;
4  import android.view.View;
5  import android.view.View.OnClickListener;
```

```
6   import android.widget.TextView;
7   import android.widget.Button;
8
9   public class MainActivity extends Activity
10  {
11      private TextView txt;
12      private Button btn;
13      public void onCreate(Bundle savedInstanceState)
14      {
15          super.onCreate(savedInstanceState);
16          setContentView(R.layout.activity_main);
17          txt=(TextView) findViewById(R.id.textView1);    ← 与界面布局文件中的相关组件建立关联
18          btn=(Button)findViewById(R.id.button1);
19          btn.setOnClickListener(new mClick());    ← 注册监听接口
20      }
21      class mClick implements OnClickListener    ← 定义实现监听接口的内部类
22      {
23          public void onClick(View v)
24          {
25              MainActivity.this.setTitle("改变标题");
26              txt.setText(R.string.newStr);
27          }
28      }
29  }
```

（3）设计资源文件 strings.xml，其代码如下：

```
1   <?xml version="1.0" encoding="utf-8"?>
2   <resources>
3       <string name="hello">Hello World, 这是 Ex03_02 的界面!</string>
4       <string name="app_name">Ex03_02</string>
5       <string name="button">单击我! </string>
6       <string name="newStr">改变了文本标签的内容</string>
7   </resources>
```

【例 3-3】编写程序，实现单击按钮改变文本标签的文字及背景颜色的功能，如图 3.5 所示。

单击按钮前　　　　　　　　　单击按钮后

图 3.5　单击按钮后，文本标签的文字及背景颜色发生变化

本例题涉及颜色定义，Android 系统在 android.graphics.Color 中定义了 12 种常见的颜色常数，其颜色常数见表 3-5。

表 3-5 常见的颜色常数

颜 色 常 数	十六进制数色码	意　　义
Color.BLACK	0xff000000	黑色
Color.BLUE	0xff00ff00	蓝色
Color.CYAN	0xff00ffff	青绿色
Color.DKGRAY	0xff444444	灰黑色
Color.GRAY	0xff888888	灰色
Color.GREEN	0xff0000ff	绿色
Color.LTGRAY	0xffcccccc	浅灰色
Color.MAGENTA	0xffff00ff	红紫色
Color.RED	0xffff0000	红色
Color.TRANSPARENT	0x00ffffff	透明
Color.WHITE	0xffffffff	白色
Color.YELLOW	0xffffff00	黄色

创建名为 Ex03_03 的新项目，包名为 com.ex03_03。

（1）设计界面布局文件 activity_main.xml。

在 XML 文件中表示颜色的方法有多种。

- #RGB：用 3 位十六进制数分别表示红、绿、蓝颜色。
- #ARGB：用 4 位十六进制数分别表示透明度，以及红、绿、蓝颜色。
- #RRGGBB：用 6 位十六进制数分别表示红、绿、蓝颜色。
- #AARRGGBB：用 8 位十六进制数分别表示透明度，以及红、绿、蓝颜色。

下面程序是用 8 位十六进制数表示透明度，以及红、绿、蓝颜色。

```
1   <?xml version="1.0" encoding="utf-8"?>
2   <LinearLayout xmlns:android="http://schemas.android.com/apk/res/android"
3       android:layout_width="fill_parent"
4       android:layout_height="fill_parent"
5       android:background="#ff7f7c"
6       android:orientation="vertical" >
7       <TextView
8           android:id="@+id/textView1"
9           android:layout_width="fill_parent"
10          android:layout_height="wrap_content"
11          android:textColor="#ff000000"     ◀── 采用 8 位十六进制数表示颜色
12          android:text="@string/hello" />
13      <Button
14          android:id="@+id/button1"
```

```
15        android:layout_width="wrap_content"
16        android:layout_height="wrap_content"
17        android:text="@string/button" />
18 </LinearLayout>
```

（2）设计控制文件 MainActivity.java，其代码如下：

```
1  package com.ex03_03;
2  import android.app.Activity;
3  import android.graphics.Color;
4  import android.os.Bundle;
5  import android.view.View;
6  import android.view.View.OnClickListener;
7  import android.widget.Button;
8  import android.widget.TextView;
9
10 public class MainActivity extends Activity
11 {
12     /** Called when the activity is first created. */
13     private TextView txt;
14     private Button btn;
15 @Override
16     public void onCreate(Bundle savedInstanceState)
17     {
18         super.onCreate(savedInstanceState);
19         setContentView(R.layout.activity_main);
20         btn=(Button)findViewById(R.id.button1);    ← 与界面布局文件中的相关组件建立关联
21         txt=(TextView)findViewById(R.id.textView1);
22         btn.setOnClickListener(new click());    ← 注册监听接口
23     }
24     class click implements OnClickListener    ← 定义实现监听接口的内部类
25     {
26        public void onClick(View v)
27         {
28            int BLACK=0xffcccccc;
29            txt.setText("改变了文字及背景颜色");
30            txt.setTextColor(Color.YELLOW);    ← 采用颜色常数设置文字颜色
31            txt.setBackgroundColor(BLACK);    ← 设置文本标签的背景颜色
32        }
33     }
34 }
```

（3）设计资源文件 strings.xml，其代码如下：

```
1  <?xml version="1.0" encoding="utf-8"?>
2  <resources>
3      <string name="hello">Hello World, MainActivity!</string>
4      <string name="app_name">Ex03_03</string>
5      <string name="button">单击我,改变文字背景颜色</string>
6  </resources>
```

3.3 文本编辑框

文本编辑框（EditText）用于接收用户输入的文本信息内容。文本编辑框继承于文本标签，其继承关系如图 3.6 所示。

```
android.view.View
    └ android.widget.TextView
          └ android.widget.EditText
```

图 3.6 文本编辑框的继承关系

文本编辑框主要继承文本标签的方法，其常用方法见表 3-6。

表 3-6 文本编辑框（EditText）的常用方法

方　　法	功　　能
EditText(Context context)	构造方法，创建文本编辑框对象
getText()	获取文本编辑框的文本内容
setText(CharSequence text)	设置文本编辑框的文本内容

其常用的 XML 文件元素属性见表 3-7。

表 3-7 文本编辑框（EditText）常用的 XML 文件元素属性

元 素 属 性	说　　明
android:editable	设置是否可编辑，其值为 true 或 false
android:numeric	设置 TextView 只能输入数字，其参数默认值为 false
android:password	设置密码输入，字符显示为圆点，其值为 true 或 false
android:phoneNumber	设置只能输入电话号码，其值为 true 或 false

定义框 EditText 元素的 android:numeric 属性，其取值只能是下列常量（可由"｜"连接多个常量）。

- integer：可以输入数值。
- signed：可以输入带符号的数值。
- decimal：可以输入带小数点的数值。

【例 3-4】设计一个密码验证程序，其运行界面如图 3.7 所示。

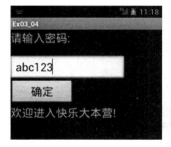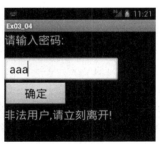

图 3.7 文本编辑框

创建名为 Ex03_04 的新项目，包名为 com.ex03_04。

（1）设计界面布局文件 activity_main.xml。在界面布局中，设置一个编辑框，用于输入密码，再设置一个按钮，判断密码是否正确，设置两个文本标签，其中一个显示提示信息"请输入密码"，另一个显示密码正确与否。其代码如下：

```xml
1  <?xml version="1.0" encoding="utf-8"?>
2  <LinearLayout xmlns:android="http://schemas.android.com/apk/res/android"
3      android:layout_width="fill_parent"
4      android:layout_height="fill_parent"
5      android:orientation="vertical" >
   <!--建立一个 TextView -->
6  <TextView
7      android:id="@+id/myTextView01"
8      android:layout_width="fill_parent"
9      android:layout_height="41px"
10     android:layout_x="33px"
11     android:layout_y="106px"
12     android:text="请输入密码:"
13     android:textSize="24sp"
14  />
    <!--建立一个 EditText -->
15 <EditText
16     android:id="@+id/myEditText"
17     android:layout_width="180px"
18     android:layout_height="wrap_content"
19     android:layout_x="29px"
20     android:layout_y="33px"
21     android:inputType="text"
22     android:textSize="24sp" />
    <!--建立一个 Button -->
23 <Button
24     android:id="@+id/myButton"
25     android:layout_width="100px"
26     android:layout_height="wrap_content"
27     android:text="确定"
28     android:textSize="24sp"
29  />
    <!--建立一个 TextView -->
30 <TextView
31     android:id="@+id/myTextView02"
32     android:layout_width="180px"
33     android:layout_height="41px"
34     android:layout_x="33px"
35     android:layout_y="106px"
```

```
36        android:textSize="24sp"
37    />
38 </LinearLayout>
```

（2）设计控制文件 MainActivity.java。在控制文件 MainActivity.java 中，主要是设计按钮的监听事件，当单击按钮后，从文本编辑框中获取输入的文本内容，与密码"abc123"进行比较。其代码如下：

```
1  package com.ex03_04;
2  import android.app.Activity;
3  import android.os.Bundle;
4  import android.view.View;
5  import android.view.View.OnClickListener;
6  import android.widget.EditText;
7  import android.widget.TextView;
8  import android.widget.Button;
9  public class MainActivity extends Activity
10 {
11     private EditText edit;
12     private TextView txt1,txt2;
13     private Button  mButton01;
14     @Override
15     public void onCreate(Bundle savedInstanceState)
16     {
17         super.onCreate(savedInstanceState);
18         setContentView(R.layout.activity_main);
19         txt1=(TextView)findViewById(R.id.myTextView01);     ← 与界面布局文件中的相关组件建立关联
20         txt2=(TextView)findViewById(R.id.myTextView02);
21         edit=(EditText)findViewById(R.id.myEditText);
22         mButton01=(Button)findViewById(R.id.myButton);
23         mButton01.setOnClickListener(new mClick());
24     }
25     class mClick implements OnClickListener     ← 定义实现监听接口的内部类
26     {
27         public void onClick(View v)
28         {
29             String passwd;
30             passwd=edit.getText().toString();     ← 获取文本编辑框中的文本内容
31             if(passwd.equals("abc123"))           ← 用 equals()方法比较两个字符串是否相等
32                 txt2.setText("欢迎进入快乐大本营!");
33             else
34                 txt2.setText("非法用户,请立刻离开!");
35         }
```

```
36    }
37 }
```

3.4　Android 布局管理

Android 系统按照 MVC（Model-View-Controller）设计模式，将应用程序的界面设计与功能控制设计分离，从而可以单独地修改用户界面，而不需要去修改程序代码。应用程序的用户界面通过 XML 定义组件布局来实现。

Android 系统的布局管理是指在 XML 布局文件中设置组件的大小、间距、排列及对齐方式等。Android 系统中常见的布局方式有 5 种，它们分别是 LinearLayout、FrameLayout、TableLayout、RelativeLayout、AbsoluteLayout。

3.4.1　布局文件的规范与重要属性

1. 布局文件的规范

Android 系统应用程序的 XML 布局文件有以下规范：

（1）布局文件作为应用项目的资源存放在 res\layout 目录下，其扩展名为.xml。

（2）布局文件的根结点通常是一个布局方式，在根结点内可以添加组件作为结点。

（3）布局文件的根结点必须包含一个命名空间：

```
xmlns:android="http://schemas.android.com/apk/res/android"
```

（4）如果要在实现控制功能的 Java 程序中控制界面中的组件，则必须为界面布局文件中的组件定义一个 ID，其定义格式为：

```
android:id="@+id/<组件 ID>"
```

2. 布局文件的重要属性

在一个界面布局中会有很多元素，这些元素的大小和位置由其属性决定。下面简述布局文件中的几个重要属性。

1）设置组件大小的属性
- wrap_content：根据组件内容的大小来决定组件的大小。
- fill_parent（或 match_parent）：使组件填充父组件容器的所有空间。

2）设置组件大小的单位
- px（pixels）：像素，即屏幕上的发光点。
- dp（或 dip，即 device independent pixels）：设备独立像素，一种支持多分辨率设备的抽象单位，和硬件相关。
- sp（scaled pixels）：比例像素，设置字体大小。

3）设置组件的对齐方式

在布局文件中，由 android:gravity 属性控制组件的对齐方式，其属性值有上（top）、下（bottom）、左（left）、右（right）、水平方向居中（center_horizontal）、垂直方向居中（center_vertical）等。

3.4.2 常见的布局方式

1. 线性布局

线性布局（LinearLayout）是 Android 系统中常用的布局方式之一，它将组件按照水平或垂直方向排列。在 XML 布局文件中，由根元素 LinearLayout 来标识线性布局。

在布局文件中，由 android:orientation 属性来控制排列方向，其属性值有水平（horizontal）和垂直（vertical）两种。

- 设置线性布局为水平方向：

android:orientation="horizontal"

- 设置线性布局为垂直方向：

android:orientation="vertical"

视频演示

【例 3-5】线性布局应用示例。

创建名为 Ex03_05 的新项目，包名为 com.ex03_05。生成项目框架后，修改界面布局文件 activity_main.xml 的代码如下：

```
1  <?xml version="1.0" encoding="utf-8"?>
2  <LinearLayout xmlns:android="http://schemas.android.com/apk/res/android"
3      android:layout_width="fill_parent"
4      android:layout_height="fill_parent"
5      android:orientation="vertical" >
6      <!-- android:orientation="horizontal" -->
7      <Button
8          android:id="@+id/mButton1"
9          android:layout_width="60px"
10         android:layout_height="wrap_content"
11         android:text="按钮 1" />
12     <Button
13         android:id="@+id/mButton2"
14         android:layout_width="60px"
15         android:layout_height="wrap_content"
16         android:text="按钮 2" />
17     <Button
18         android:id="@+id/mButton3"
19         android:layout_width="60px"
20         android:layout_height="wrap_content"
21         android:text="按钮 3" />
22     <Button
23         android:id="@+id/mButton4"
24         android:layout_width="60px"
25         android:layout_height="wrap_content"
26         android:text="按钮 4" />
27 </LinearLayout>
```

程序的运行结果如图 3.8（a）所示。如果将代码中的第 5 行 android:orientation="vertical"（垂直方向的线性布局）更改为 android:orientation="horizontal"（水平方向的线性布局），则运行结果如图 3.8（b）所示。

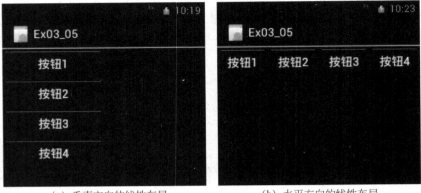

(a) 垂直方向的线性布局　　　　　(b) 水平方向的线性布局

图 3.8　线性布局示例

2. 帧布局

帧布局 FrameLayout 是将组件放置到左上角位置，当添加多个组件时，后面的组件将遮盖之前的组件。在 XML 布局文件中，由根元素 FrameLayout 来标识帧布局。

【例 3-6】帧布局应用示例。

创建名为 Ex03_06 的新项目，包名为 com.ex03_06。生成项目框架后，将事先准备的图像文件 img.png 复制到 res\drawable-hdpi 目录下。

视频演示

（1）设计界面布局文件 activity_main.xml，其代码如下：

```
1  <?xml version="1.0" encoding="utf-8"?>
2  <FrameLayout
3    xmlns:android="http://schemas.android.com/apk/res/android"
4      android:layout_width="fill_parent"
5      android:layout_height="fill_parent">
6    <ImageView
7        android:id="@+id/mImageView"
8        android:layout_width="60px"
9        android:layout_height="wrap_content"
10   />
11   <TextView
12       android:layout_width="wrap_content"
13       android:layout_height="wrap_content"
14       android:text="快乐大本营"
15       android:textSize="18sp"
16   />
17 </FrameLayout>
```

（2）设计控制文件 MainActivity.java，其代码如下：

```
1   package com.ex03_06;
2   import android.app.Activity;
3   import android.os.Bundle;
4   import android.widget.ImageView;
5   public class MainActivity extends Activity
6   {
7       ImageView imageview;
8       @Override
9       public void onCreate(Bundle savedInstanceState)
10      {
11          super.onCreate(savedInstanceState);
12          setContentView(R.layout.activity_main);
13          imageview=(ImageView) this.findViewById(R.id.mImageView);
14          imageview.setImageResource(R.drawable.img);
15      }
16  }
```

程序运行结果如图 3.9 所示，可见在界面布局文件中添加的文本框组件遮挡了之前的图像组件。

3．表格布局

表格布局（TableLayout）是将页面划分成由行、列构成的单元格。在 XML 布局文件中，由根元素 TableLayout 来标识表格布局。

表格的列数由 android:shrinkColumns 定义。例如，android:shrinkColumns = "0, 1, 2"，表示表格为 3 列，其列编号为第 1、2、3。

表格的行由<TableRow></TableRow>定义。组件放置到哪一列，由 android:layout_column 指定列编号。

【例 3-7】表格布局应用示例。设计一个 3 行 4 列的表格布局，组件安排如图 3.10 所示。

图 3.9　帧布局示例

图 3.10　3 行 4 列的表格布局

视频演示

创建名为 Ex03_07 的新项目，包名为 com.ex03_07。生成项目框架后，将准备好的图像文件 img1.png、img2.png、img3.png、img4.png、img5.png 复制到 res\drawable-hdpi 目录下。

（1）设计表格的界面布局文件 activity_main.xml。在图 3.10 所示的界面布局中，由于有显示图片的空白单元格，这时，可以使用文本标签组件将其文字内容设置为空，这样显示出来的就是空白的单元格了。该文件的代码如下：

```xml
1  <?xml version="1.0" encoding="utf-8"?>
2  <TableLayout xmlns:android="http://schemas.android.com/apk/res/android"
3      android:layout_width="fill_parent"
4      android:layout_height="fill_parent">
5  <TableRow>    <!-- 第1行 -->
6  <ImageView  android:id="@+id/mImageView1"          ← 第1列
7      android:layout_width="wrap_content "
8      android:layout_height="wrap_content"
9      android:src="@drawable/img1" />
10 <ImageView  android:id="@+id/mImageView2"          ← 第2列
11     android:layout_width="wrap_content "
12     android:layout_height="wrap_content"
13     android:src="@drawable/img2" />
14 </TableRow>
15 <TableRow>    <!-- 第2行 -->
16 <TextView
17     android:id="@+id/textView1"                   ← 第1列空白单元格
18     android:layout_width="wrap_content"
19     android:layout_height="wrap_content" />
20 <ImageView  android:id="@+id/mImageView3"          ← 第2列
21     android:layout_width="wrap_content "
22     android:layout_height="wrap_content"
23     android:src="@drawable/img3" />
24 <ImageView  android:id="@+id/mImageView4"          ← 第3列
25     android:layout_width=" wrap_content "
26     android:layout_height="wrap_content"
27     android:src="@drawable/img4" />
28 </TableRow>
29 <TableRow>    <!-- 第3行 -->
30 <TextView
31     android:id="@+id/textView2"                   ← 第1列空白单元格
32     android:layout_width="wrap_content"
33     android:layout_height="wrap_content" />
34 <TextView
35     android:id="@+id/textView3"                   ← 第2列空白单元格
36     android:layout_width="wrap_content"
37     android:layout_height="wrap_content" />
38 <TextView
39     android:id="@+id/textView4"                   ← 第3列空白单元格
40     android:layout_width="wrap_content"
41     android:layout_height="wrap_content" />
42 <ImageView  android:id="@+id/mImageView5"          ← 第4列
43     android:layout_width="wrap_content "
44     android:layout_height="wrap_content"
45     android:src="@drawable/img5" />
46 </TableRow>
47 </TableLayout>
```

（2）设计控制文件 MainActivity.java，其代码如下：

```
1  package com.ex03_07;
2  import android.app.Activity;
3  import android.os.Bundle;
4  import android.widget.ImageView;
5  public class MainActivity extends Activity
6  {
7      ImageView img1, img2, img3, img4, img5;
8      @Override
9      public void onCreate(Bundle savedInstanceState)
10     {
11         super.onCreate(savedInstanceState);
12         setContentView(R.layout.activity_main);
13         img1=(ImageView)this.findViewById(R.id.mImageView1);
14         img2=(ImageView)this.findViewById(R.id.mImageView2);
15         img3=(ImageView)this.findViewById(R.id.mImageView3);
16         img4=(ImageView)this.findViewById(R.id.mImageView4);
17         img5=(ImageView)this.findViewById(R.id.mImageView5);
18         img1.setImageResource(R.drawable.img1);
19         img2.setImageResource(R.drawable.img2);
20         img3.setImageResource(R.drawable.img3);
21         img4.setImageResource(R.drawable.img4);
22         img5.setImageResource(R.drawable.img5);
23     }
24 }
```

4. 相对布局

相对布局（RelativeLayout）是采用相对其他组件的位置的布局方式。在相对布局中，通过指定 id 关联其他组件，以右对齐、上对齐、下对齐或居中对齐等方式来排列组件。

在 XML 布局文件中，由根元素 RelativeLayout 来标识相对布局。相对布局由于属性较多，下面简单介绍几种常用属性。

设置该控件与父元素右对齐：

android:layout_alignParentRight="true"

设置该控件在 id 为 re_edit_0 控件的下方：

android:layout_below="@id/re_edit_0"

设置该控件在 id 为 re_image_0 控件的左边：

android:layout_toLeftOf="@id/re_iamge_0"

设置当前控件与 id 为 name 控件的上方对齐：

android:layout_alignTop="@id/name"

设置偏移的像素值：

```
android:layout_marginRight="30dip"
```

该布局方式的属性较多，下面简单归纳一下：

- 第一类：属性值为 true 或 false。

```
android:layout_centerHrizontal
android:layout_centerVertical
android:layout_centerInparent
android:layout_alignParentBottom
android:layout_alignParentLeft
android:layout_alignParentRight
android:layout_alignParentTop
android:layout_alignWithParentIfMissing
```

- 第二类：属性值必须为 id 的引用名"@id/id-name"。

```
android:layout_below
android:layout_above
android:layout_toLeftOf
android:layout_toRightOf
android:layout_alignTop
```

- 第三类：属性值为具体的像素值，如 30dp。

```
android:layout_marginBottom
android:layout_marginLeft
android:layout_marginRight
android:layout_marginTop
```

【例3-8】应用相对布局设计一个组件排列如图 3.11 所示的应用程序。

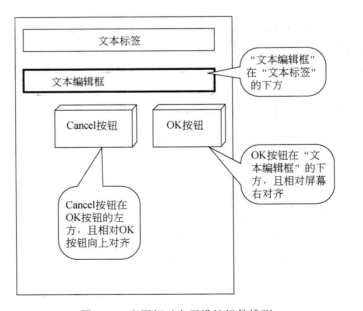

图 3.11 应用相对布局设计组件排列

创建名为 Ex03_08 的新项目，包名为 com.ex03_08。生成项目框架后，修改界面布局文件 activity_main.xml 的代码如下：

```
1  <?xml version="1.0" encoding="utf-8"?>
2  <RelativeLayout xmlns:android="http://schemas.android.com/apk/res/android"
3      android:layout_width="fill_parent"
4      android:layout_height="fill_parent">
5      <TextView
6          android:id="@+id/label"
7          android:layout_width="fill_parent"
8          android:layout_height="wrap_content"
9          android:textSize="24sp"
10         android:text="相对布局"/>
11     <EditText
12         android:id="@+id/edit"
13         android:layout_width="fill_parent"
14         android:layout_height="wrap_content"
15         android:background="@android:drawable/editbox_background"
16         android:layout_below="@id/label"/>      ← 在文本标签的下方
17     <Button
18         android:id="@+id/ok"
19         android:layout_width="wrap_content"
20         android:layout_height="wrap_content"
21         android:layout_below="@id/edit"          ← 在文本编辑框的下方
22         android:layout_alignParentRight="true"   ← 与父容器右对齐
23         android:layout_marginLeft="10dip"
24         android:text="OK" />
25     <Button
26         android:layout_width="wrap_content"
27         android:layout_height="wrap_content"
28         android:layout_toLeftOf="@id/ok"         ← 在 OK 按钮的左方
29         android:layout_alignTop="@id/ok"         ← 与 OK 按钮顶部对齐
30         android:text="Cancel" />
31 </RelativeLayout>
```

程序的运行结果如图 3.12 所示。

5. 绝对布局

绝对布局（AbsoluteLayout）是在界面布局文件中指定组件在屏幕上的坐标位置。在 XML 布局文件中，由根元素 AbsoluteLayout 来标识绝对布局。

如果要正确应用绝对布局安排组件的位置，需要了解 Android 图形界面的坐标系统。

图 3.12 相对布局

在一个二维的 Android 图形界面坐标系中,该坐标的原点在组件的左上角,坐标的单位是像素。X 轴在水平方向上从左至右,Y 轴在垂直方向上从上向下,如图 3.13 所示。

【例 3-9】 绝对布局示例。

设一个文本标签在屏幕上的坐标为(40,150),修改界面布局文件 activity_main.xml 的代码如下:

```
1  <?xml version="1.0" encoding="utf-8"?>
2  <AbsoluteLayout xmlns:android="http://schemas.android.com/apk/res/android"
3      android:layout_width="fill_parent"
4      android:layout_height="fill_parent"
5      android:orientation="vertical" >
6      <TextView
7          android:layout_width="fill_parent"
8          android:layout_height="wrap_content"
9          android:layout_x="40dp"
10         android:layout_y="150dp"
11         android:text="欢迎进入 Android 世界!" />
12 </AbsoluteLayout>
```

程序的运行结果如图 3.14 所示。

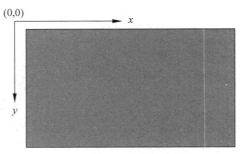

图 3.13 坐标系统

图 3.14 绝对布局

3.5 进度条和选项按钮

3.5.1 进度条

进度条(ProgressBar)能以形象地图示方式直观地显示某个过程的进度。进度条的常用属性和方法见表 3-8。

表 3-8 进度条(ProgressBar)的常用属性和方法

属 性	方 法	功 能
android:max	setMax(int max)	设置进度条的变化范围为 0~max
android:progress	setProgress(int progress)	设置进度条的当前值(初始值)
android:progressby	incrementProgressBy(int diff)	设置进度条的变化步长值

【例3-10】进度条应用示例。

在界面设计中安排一个进度条组件，并设置两个按钮，用于控制进度条的进度变化，如图3.15所示。

程序设计步骤：
（1）在界面布局文件中声明ProgressBar。
（2）在Activity中获得ProgressBar实例。
（3）调用ProgressBar的incrementProgressBy()方法增加或减少进度。

程序代码：

（1）设计界面布局文件activity_main.xml的代码如下：

图3.15 进度条进度控制

```
1  <?xml version="1.0" encoding="utf-8"?>
2  <LinearLayout xmlns:android="http://schemas.android.com/apk/res/android"
3      android:layout_width="fill_parent"
4      android:layout_height="fill_parent"
5      android:orientation="vertical" >
6      <ProgressBar
7          android:id="@+id/ProgressBar01"
8          style="@android:style/Widget.ProgressBar.Horizontal"
9          android:layout_width="250dp"
10         android:layout_height="wrap_content"
11         android:max="200"
12         android:progress="50" >
13     </ProgressBar>
14     <Button
15         android:id="@+id/button1"
16         android:layout_width="wrap_content"
17         android:layout_height="wrap_content"
18         android:text="@string/btn1" />
19     <Button
20         android:id="@+id/button2"
21         android:layout_width="wrap_content"
22         android:layout_height="wrap_content"
23         android:text="@string/btn2" />
24 </LinearLayout>
```

（2）在控制文件MainActivity.java中添加按钮的事件处理代码：

```
1  package com.ex03_10;
2  import android.app.Activity;
3  import android.os.Bundle;
4  import android.view.View;
```

```
5  import android.view.View.OnClickListener;
6  import android.widget.Button;
7  import android.widget.ProgressBar;

8  public class MainActivity extends Activity
9  {
10 ProgressBar progressBar;
11 Button btn1,btn2;
12   @Override
13 public void onCreate(Bundle savedInstanceState)
14 {
15   super.onCreate(savedInstanceState);
16   setContentView(R.layout.activity_main);
17   progressBar=(ProgressBar)findViewById(R.id.ProgressBar01);
18   btn1=(Button)findViewById(R.id.button1);           ┐  与界面布局文件中的
19   btn2=(Button)findViewById(R.id.button2);           ┘  相关组件建立关联
20   btn1.setOnClickListener(new click1());             ┐  注册按钮的监听器
21   btn2.setOnClickListener(new click2());             ┘
22 }
23 class click1 implements OnClickListener
24 {
25     public void onClick(View v)
26       { progressBar.incrementProgressBy(5);  }       ◄── 增加进度
27 }
28 class click2 implements OnClickListener
29 {
30     public void onClick(View v)
31       { progressBar.incrementProgressBy(-5);  }      ◄── 减少进度
32 }
33 }
```

3.5.2 选项按钮

1. 复选框

复选框（CheckBox）用于多项选择，用户可以一次性选择多个选项。复选框是按钮的子类，其属性和方法继承于按钮。复选框的常用方法见表3-9。

表 3-9 复选框（CheckBox）的常用方法

方　　法	功　　能
isChecked()	判断选项是否被选中
getText()	获取复选框的文本内容

【例 3-11】复选框应用示例。

在界面设计中,安排 3 个复选框和一个普通按钮,选择选项后,单击按钮,在文本标签中显示所选中的选项文本内容,如图 3.16 所示。

视频演示

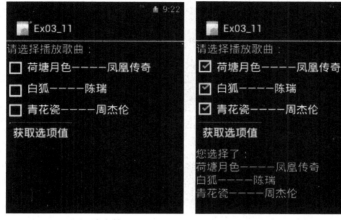

(a) 选中前　　　　　　(b) 选中后

图 3.16　复选框应用示例

程序设计步骤:

(1) 在界面布局文件中声明复选框。

(2) 在 Activity 中获得复选框实例。

(3) 调用 CheckBox 的 isChecked()方法判断该选项是否被选中。如果被选中,则调用 getText()方法获取选项的文本内容。

程序代码:

(1) 设计界面布局文件 activity_main.xml 的代码如下:

```
1   <?xml version="1.0" encoding="utf-8"?>
2   <LinearLayout xmlns:android="http://schemas.android.com/apk/res/android"
3       android:layout_width="fill_parent"
4       android:layout_height="fill_parent"
5       android:orientation="vertical" >
6       <TextView
7           android:layout_width="fill_parent"
8           android:layout_height="wrap_content"
9           android:text="@string/hello"
10          android:textSize="20sp"/>
11      <CheckBox
12          android:id="@+id/check1"
13          android:layout_width="fill_parent"
14          android:layout_height="wrap_content"
15          android:textSize="20sp"
```

```
16         android:text="@string/one" />
17     <CheckBox
18         android:id="@+id/check2"
19         android:layout_width="fill_parent"
20         android:layout_height="wrap_content"
21         android:textSize="20sp"
22         android:text="@string/two" />
23     <CheckBox
24         android:id="@+id/check3"
25         android:layout_width="fill_parent"
26         android:layout_height="wrap_content"
27         android:textSize="20sp"
28         android:text="@string/three" />
29     <Button
30         android:id="@+id/button"
31         android:layout_width="wrap_content"
32         android:layout_height="wrap_content"
33         android:textSize="20sp"
34         android:text="@string/btn" />
35     <TextView
36         android:id="@+id/textView2"
37         android:layout_width="fill_parent"
38         android:layout_height="wrap_content"
39         android:text=""
40         android:textSize="20sp"/>
41 </LinearLayout>
```

（2）在 strings.xml 文件中设置要使用的字符串，代码如下：

```
1 <?xml version="1.0" encoding="utf-8"?>
2 <resources>
3     <string name="hello">请选择播放歌曲: </string>
4     <string name="app_name">Ex03_11</string>
5     <string name="one">荷塘月色----凤凰传奇</string>
6     <string name="two">白狐----陈瑞</string>
7     <string name="three">青花瓷----周杰伦</string>
8     <string name="btn">获取选项值</string>
9 </resources>
```

（3）设计控制文件 MainActivity.java。在控制文件 MainActivity.java 中，建立组件与界面布局文件中相关组件的关联，并编写按钮的事件处理代码：

```
1 package com.ex03_11;
```

```
2   import android.app.Activity;
3   import android.os.Bundle;
4   import android.view.View;
5   import android.view.View.OnClickListener;
6   import android.widget.Button;
7   import android.widget.CheckBox;
8   import android.widget.TextView;
9   public class MainActivity extends Activity
10  {
11      CheckBox ch1,ch2,ch3;
12      Button okBtn;
13      TextView txt;
14      @Override
15      public void onCreate(Bundle savedInstanceState)
16      {
17          super.onCreate(savedInstanceState);
18          setContentView(R.layout.activity_main);
19          ch1=(CheckBox)findViewById(R.id.check1);
20          ch2=(CheckBox)findViewById(R.id.check2);
21          ch3=(CheckBox)findViewById(R.id.check3);
22          okBtn=(Button)findViewById(R.id.button);
23          txt=(TextView)findViewById(R.id.textView2);
24          okBtn.setOnClickListener(new click());
25      }
26      class click implements OnClickListener
27      {
28          public void onClick(View v)
29          {
30            String str="";
31            if(ch1.isChecked()) str=str+"\n"+ch1.getText();
32            if(ch2.isChecked()) str=str+"\n"+ch2.getText();
33            if(ch3.isChecked()) str=str+"\n"+ch3.getText();
34            txt.setText("您选择了: "+str);
35          }
36  }
37  }
```

（第19-23行注释：与界面布局文件中的相关组件建立关联）

2. 单选组件与单选按钮

单选组件（RadioGroup）用于多项选择中只允许任选其中一项的情形。单选组件由一组单选按钮（RadioButton）组成。单选按钮是按钮的子类。单选按钮的常用方法见表3-10。

表 3-10　单选按钮（RadioButton）的常用方法

方　　法	功　　能
isChecked();	判断选项是否被选中
getText();	获取单选按钮的文本内容

【例 3-12】单选按钮应用示例。

在界面设计中，安排两个单选按钮、一个文本编辑框和一个普通按钮，选择选项后，单击按钮，在文本标签中显示文本编辑框及选中选项的文本内容，如图 3.17 所示。

视频演示

程序设计步骤：

（1）在界面布局文件中声明单选组件和单选按钮。

（2）在 Activity 中获得单选按钮实例。

（3）调用 RadioButton 的 isChecked()方法判断该选项是否被选中。如果被选中，则调用 getText()方法获取选项的文本内容。

图 3.17　单选按钮示例

程序代码：

（1）设计界面布局文件 activity_main.xml 的代码如下：

```
1  <?xml version="1.0" encoding="utf-8"?>
2  <LinearLayout xmlns:android="http://schemas.
   android.com/apk/res/android"
3      android:layout_width="fill_parent"
4      android:layout_height="fill_parent"
5      android:orientation="vertical" >
6   <TextView
7      android:layout_width="fill_parent"
8      android:layout_height="wrap_content"
9      android:textSize="20sp"
10     android:text="@string/hello" />
11  <EditText
12     android:id="@+id/edit1"
13     android:layout_width="fill_parent"
14     android:layout_height="wrap_content"
15     android:inputType="text"
16     android:textSize="20sp" />
17  <RadioGroup
18     android:layout_width="fill_parent"
19     android:layout_height="wrap_content">
20  <RadioButton
21     android:id="@+id/boy01"
22     android:text="@string/boy"/>
23  <RadioButton
24     android:id="@+id/girl01"
```

```
25          android:text="@string/girl" />
26      </RadioGroup>
27      <Button
28          android:id="@+id/myButton"
29          android:layout_width="wrap_content"
30          android:layout_height="wrap_content"
31          android:text="确定"
32          android:textSize="20sp"
33      />
34      <TextView
35          android:id="@+id/text02"
36          android:layout_width="fill_parent"
37          android:layout_height="wrap_content"
38          android:textSize="20sp"
39      />
40  </LinearLayout>
```

（2）在 strings.xml 文件中设置要使用的字符串，代码如下：

```
1   <?xml version="1.0" encoding="utf-8"?>
2   <resources>
3       <string name="hello">请输入您的姓名：</string>
4       <string name="app_name">Ex03_12</string>
5       <string name="boy">男</string>
6       <string name="girl">女</string>
7   </resources>
```

（3）设计控制文件 MainActivity.java。在控制文件 MainActivity.java 中，建立组件与界面布局文件中相关组件的关联，并编写按钮的事件处理代码：

```
1   package com.ex03_12;
2   import android.app.Activity;
3   import android.os.Bundle;
4   import android.view.View;
5   import android.view.View.OnClickListener;
6   import android.widget.Button;
7   import android.widget.EditText;
8   import android.widget.RadioButton;
9   import android.widget.TextView;
10  public class MainActivity extends Activity
11  {
12   Button okBtn;
13   EditText edit;
14   TextView txt;
15   RadioButton r1, r2;
16   @Override
```

```
17  public void onCreate(Bundle savedInstanceState)
18  {
19      super.onCreate(savedInstanceState);
20      setContentView(R.layout.activity_main);
21      edit=(EditText)findViewById(R.id.edit1);
22      okBtn=(Button)findViewById(R.id.myButton);
23      txt=(TextView)findViewById(R.id.text02);
24      r1=(RadioButton)findViewById(R.id.boy01);
25      r2=(RadioButton)findViewById(R.id.girl01);
26      okBtn.setOnClickListener(new mClick());
27  }
28  class mClick implements OnClickListener
29  {
30    public void onClick(View v)
31    {
32      CharSequence str="", name="";
33      name=edit.getText();
34      if(r1.isChecked())          ← 第1个单选按钮被选中
35          str=r1.getText();
36      if(r2.isChecked())          ← 第2个单选按钮被选中
37          str=r2.getText();
38      txt.setText("您输入的信息为：\n 姓名 "+name+"\t 性别 "+str);
39    }
40  }
41 }
```

21—25 行右侧注释：与界面布局文件中的相关组件建立关联

3.6 图像显示与画廊组件

3.6.1 图像显示 ImageView 类

ImageView 类用于显示图片或图标等图像资源，并提供图像缩放及着色（渲染）等图像处理功能。

ImageView 类的常用属性和对应方法见表 3-11。

表 3-11 ImageView 类的常用属性和对应方法

元素属性	对应方法	说明
android:maxHeight	setMaxHeight(int)	为显示图像提供最大高度的可选参数
android:maxWidth	setMaxWidth(int)	为显示图像提供最大宽度的可选参数
android:ScaleType	setScaleType(ImageView.ScaleType)	控制图像适合 ImageView 大小的显示方式
android:src	setImageResource(int)	获取图像文件的路径

ImageView 类的 ScaleType 属性值见表 3-12。

表 3-12 ImageView 类的 ScaleType 属性值

ScaleType 属性值常量	值	说明
matrix	0	用矩阵来绘图
fitXY	1	拉伸图片（不按宽高比例）以填充 View 的宽高
fitStart	2	按比例拉伸图片，拉伸后图片的高度为 View 的高度，且显示在 View 的左边
fitCenter	3	按比例拉伸图片，拉伸后图片的高度为 View 的高度，且显示在 View 的中间
fitEnd	4	按比例拉伸图片，拉伸后图片的高度为 View 的高度，且显示在 View 的右边
center	5	按原图大小显示图片，但图片的宽高大于 View 的宽高时，截取图片中间部分显示
centerCrop	6	按比例放大原图，直至等于某边 View 的宽高显示
centerInside	7	当原图宽高等于 View 的宽高时，按原图大小居中显示，否则将原图缩放至 View 的宽高居中显示

【例 3-13】显示图片示例。

程序设计步骤：

（1）将事先准备好的图片序列 img1.jpg、img2.jpg、…、img6.jpg 复制到 res\drawable-hdpi 目录下。

（2）在布局文件中声明图像显示组件 ImageView。

（3）在 Activity 中获得相关组件实例。

（4）通过触发按钮事件，调用 OnClickListener 接口的 onClick()方法显示图像。

程序代码：

（1）设计界面布局文件 activity_main.xml。在界面设计中，安排两个按钮和一个图像显示组件 ImageView，单击按钮，可以翻阅浏览图片。布局设计如图 3.18 所示。

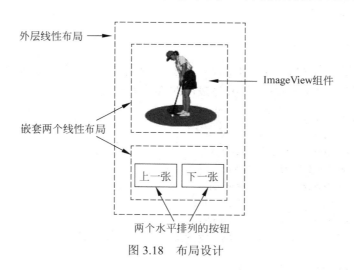

图 3.18 布局设计

其代码如下：

```xml
1  <?xml version="1.0" encoding="utf-8"?>
2  <LinearLayout xmlns:android="http://schemas.android.com/apk/res/android"
3  android:layout_width="fill_parent"
4  android:layout_height="fill_parent"
5  android:gravity="center|fill"
6  android:orientation="vertical" >
7  <LinearLayout
8  android:layout_width="fill_parent"
9  android:layout_height="wrap_content"
10 android:gravity="center" >
11 <ImageView
12 android:id="@+id/img"
13 android:layout_width="240dp"
14 android:layout_height="240dp"
15 android:layout_centerVertical="true"
16 android:src="@drawable/img1" />
17 </LinearLayout>
18   <LinearLayout
19   android:layout_width="fill_parent"
20   android:layout_height="wrap_content" >
21  <Button
22    android:id="@+id/btn_last"
23    android:layout_width="150dp"
24    android:layout_height="wrap_content"
25    android:text="上一张" />
26  <Button android:id="@+id/btn_next"
27    android:layout_width="150dp"
28    android:layout_height="wrap_content"
29    android:text="下一张"  />
30  </LinearLayout>
31 </LinearLayout>
```

（2）设计控制文件 MainActivity.java。在控制文件 MainActivity.java 中，建立组件与界面布局文件中相关组件的关联，并编写按钮的事件处理代码：

```java
1  package com.ex03_13;
2  import android.app.Activity;
3  import android.os.Bundle;
4  import android.view.View;
5  import android.view.View.OnClickListener;
6  import android.widget.Button;
7  import android.widget.ImageView;
8  public class MainActivity extends Activity {
9  ImageView img;
10 Button btn_last, btn_next;
```

```
11  //存放图片id的int数组
12  private int[] imgs={
13  R.drawable.img1,
14  R.drawable.img2,
15  R.drawable.img3,
16  R.drawable.img4,
17  R.drawable.img5,
18  R.drawable.img6 };
19  int index=1;
20  @Override
21  public void onCreate(Bundle savedInstanceState) {
22  super.onCreate(savedInstanceState);
23  setContentView(R.layout.activity_main);
24  img=(ImageView)findViewById(R.id.img);
25  btn_last=(Button)findViewById(R.id.btn_last);
26  btn_next=(Button)findViewById(R.id.btn_next);
27  btn_last.setOnClickListener(new mClick());
28  btn_next.setOnClickListener(new mClick());
29  }
30  class mClick implements OnClickListener
31  {
32    public void onClick(View v)
33    {
34     if(v==btn_last)
35     {
36      if(index>0 && index<imgs.length)
37      {
38        index--;
39        img.setImageResource(imgs[index]);
40      } else {index=imgs.length+1; }
41     }
42     if(v==btn_next)
43     {
44        if(index>0&&index<imgs.length-1)
45        {
46          index++;
47          img.setImageResource(imgs[index]);
48        }else {index=imgs.length-1;      }
49     }
50    }
51  }
52  }
```

程序的运行结果如图3.19所示。

图 3.19 图像显示示例

3.6.2 画廊组件 Gallery 与图片切换器 ImageSwitcher

Gallery 是 Android 中控制图片展示的组件，它可以横向显示一列图像。Gallery 的常用属性及方法见表 3-13。

表 3-13 Gallery 的常用属性及方法

元素属性	对应方法	说 明
android:spacing	setSpacing(int)	设置图片之间的间距，以像素为单位
android:unselectedAlpha	setUnselectedAlpha(float)	设置未选中图片的透明度（Alpha）
android:animationDuration	set AnimationDuration(int)	设置布局变化时动画转换所需的时间（毫秒级），仅在动画开始时计时
	onTouchEvent(MotionEvent event)	触摸屏幕时触发 MotionEvent 事件
	onDown(MotionEvent e)	按下屏幕时触发 MotionEvent 事件

Gallery 经常与图片切换器 ImageSwitcher 配合使用，用图片切换器 ImageSwitcher 展示图片效果。使用 ImageSwitcher 时必须用 ViewFactory 接口的 makeView()方法创建视图。ImageSwitcher 的常用方法见表 3-14。

表 3-14 ImageSwitcher 的常用方法

方 法	说 明
setInAnimation (Animation inAnimation)	设置动画对象进入屏幕的方式
setOutAnimation(Animation outAnimation)	设置动画对象退出屏幕的方式
setImageResource(int resid)	设置显示的初始图片
showNext()	显示下一个视图
showPrevious()	显示前一个视图

【例3-14】画廊展示图片示例。

在界面设计中,安排一个画廊组件 Gallery 和一个图片切换器 ImageSwitcher,单击画廊中的小图片,可以在图片切换器中显示放大的图片,如图 3.20 所示。

视频演示

图 3.20 画廊展示图片示例

程序设计步骤:

(1) 在界面布局文件中声明画廊组件 Gallery 和图片切换器 ImageSwitcher,采用表格布局。

(2) 把事先准备好的图片文件 img1.jpg、img2.jpg、…、img8.jpg 复制到项目的资源目录 res\drawable-hdpi 中,在 Activity 中创建一个图像文件数组 imgs[],其数组元素为图片文件。

(3) 在 Activity 中创建画廊组件 Gallery 和图片切换器 ImageSwitcher 组件的实例对象。

(4) 在 Activity 中创建一个实现 ViewFactory 接口的内部类,重写 makeView()方法建立 imageView 图像视图。图片切换器 ImageSwitcher 通过该图像视图显示放大的图片。

(5) 在 Activity 中创建一个 BaseAdapter 适配器,用于安排放在画廊 Gallery 中的图片文件及显示方式。

程序代码:

(1) 设计界面布局文件 activity_main.xml 的代码如下:

```
1   <?xml version="1.0" encoding="utf-8"?>
2   <TableLayout android:id="@+id/TableLayout01"
3     android:layout_width="wrap_content"
4     android:layout_height="wrap_content"
5     xmlns:android="http://schemas.android.com/apk/res/android"
6     android:layout_gravity="center">
7     <TextView
8       android:layout_width="fill_parent"
9       android:layout_height="wrap_content"
```

```
10        android:textSize="20sp"
11        android:text="@string/hello" />
12   <Gallery android:id="@+id/Gallery01"
13        android:layout_width="wrap_content"
14        android:layout_height="wrap_content"
15        android:spacing="10dp"/>
16   <ImageSwitcher android:id="@+id/ImageSwitcher01"
17        android:layout_width="wrap_content"
18        android:layout_height="wrap_content" >
19   </ ImageSwitcher>
20 /TableLayout>
```

（2）设计控制文件 MainActivity.java。在控制文件 MainActivity.java 中创建图像文件序列数组，并编写按钮的事件处理代码。通过 ViewFactory 接口建立 imageView 图像视图，并实现 OnItemSelectedListener 接口来选择图片。其代码如下：

```
1  package com.ex03_14;
2  import android.app.Activity;
3  import android.os.Bundle;
4  import android.view.View;
5  import android.view.ViewGroup;
6  import android.view.animation.AnimationUtils;
7  import android.widget.AdapterView;
8  import android.widget.BaseAdapter;
9  import android.widget.Gallery;
10 import android.widget.ImageSwitcher;
11 import android.widget.ImageView;
12 import android.widget.AdapterView.OnItemSelectedListener;
13 import android.widget.ViewSwitcher.ViewFactory;
14
15 public class MainActivity extends Activity
16 {
17     private ImageSwitcher imageSwitcher;
18      Gallery gallery;
19     private int[] imgs={
20         R.drawable.img1,
21         R.drawable.img2,
22         R.drawable.img3,
23         R.drawable.img4,
24         R.drawable.img5,
25         R.drawable.img6,
26         R.drawable.img7,
27         R.drawable.img8,
28     };
29
30   @Override
```

```
31  public void onCreate(Bundle savedInstanceState)
32  {
33    super.onCreate(savedInstanceState);
34    setContentView(R.layout.activity_main);
35    imageSwitcher=(ImageSwitcher)findViewById(R.id.ImageSwitcher01);
36    imageSwitcher.setFactory(new viewFactory());
37    imageSwitcher.setInAnimation(AnimationUtils
38        .loadAnimation(this, android.R.anim.fade_in) );     ← 设置淡入 fade_in、
39    imageSwitcher.setOutAnimation(AnimationUtils              淡出 fade_out 方式
40        .loadAnimation(this, android.R.anim.fade_out));
41    imageSwitcher.setImageResource(R.drawable.img1);  ← 设置显示的初始图片
42    gallery=(Gallery)findViewById(R.id.Gallery01);
43    gallery.setOnItemSelectedListener(     ← 设置监听，获取选择的图片
44            new onItemSelectedListener());
45    gallery.setSpacing(10);  ← 设定画廊 gallery 的图片之间的间隔（像素为单位）
46    gallery.setAdapter(new baseAdapter());  ← 设置图片文件及显示方式的适配器
47  }
48  //通过 ViewFactory 接口建立一个 imageView 图像视图
49  class viewFactory implements ViewFactory
50  {
51    @Override
52    public View makeView()
53    {
54      ImageView imageView=new ImageView(MainActivity.this);
55      imageView.setScaleType(ImageView.ScaleType.FIT_CENTER);  ← 设置显示方式
56      return imageView;
57    }
58  }
59  //实现选项监听接口,获取选择的图片
60  class onItemSelectedListener  implements OnItemSelectedListener
61  {
62    @Override
63    public void onItemSelected(AdapterView<?> parent,   ← 监听选项
64                    View view,int position, long id)
65    {
66        imageSwitcher.setImageResource(
67                  (int)gallery.getItemIdAtPosition(position));
68    }
69    @Override
70    public void onNothingSelected(AdapterView<?> arg0) { }
71  }
72  //设置一个适配器,安排放在画廊 gallery 中的图片文件及显示方式
73  class  baseAdapter extends BaseAdapter
74  {
75    //取得 gallery 内的照片数量
```

```
76    public int getCount()
77      {return imgs.length;}
78    public Object getItem(int position)
79      { return null;          }
80    //取得gallery内选择的某一张图片文件
81    public long getItemId(int position)
82      {    return imgs[position]; }
83    //将选择的图片放置在imageView,且设定显示方式为居中,大小为60×60
84    public View getView(int position, View convertView, ViewGroup parent)
85    {
86        ImageView imageView = new ImageView(MainActivity.this);
87        imageView.setImageResource(imgs[position]);
88        imageView.setScaleType(ImageView.ScaleType.FIT_CENTER);
89        imageView.setLayoutParams(new Gallery.LayoutParams(60, 60));
90        return imageView;
91    }
92  }
93 }
```

3.7 消息提示

在 Android 系统中,可以用 Toast 来显示帮助或提示消息。该提示消息以浮于应用程序之上的形式显示在屏幕上。因为它并不获得焦点,不会影响用户的其他操作,使用消息提示组件 Toast 的目的就是为了尽可能不中断用户操作,并使用户看到提供的信息内容。Toast 类的常用方法见表 3-15。

表 3-15 Toast 类的常用方法

常 用 方 法	说　　明
Toast(Context context)	Toast 的构造方法,构造一个空的 Toast 对象
makeText(Context context, CharSequence text, int duration)	以特定时长显示文本内容,参数 text 为显示的文本,参数 duration 为显示时间,较长时间取值 LENGTH_LONG,较短时间取值 LENGTH_SHORT
getView()	返回视图
setDuration(int duration)	设置存续时间
setView(View view)	设置要显示的视图
setGravity(int gravity, int xOffset, int yOffset)	设置提示信息在屏幕上的显示位置
setText(int resId)	更新 makeText()方法所设置的文本内容
show()	显示提示信息
LENGTH_LONG	提示信息显示较长时间的常量
LENGTH_SHORT	提示信息显示较短时间的常量

【例3-15】消息提示 Toast 分别按默认方式、自定义方式和带图标方式显示的示例。

将事先准备好的图标文件 icon.jpg 复制到 res\drawable-hdpi 目录下,以作提示消息的图标之用。

(1) 设计界面布局文件 activity_main.xml。在界面设计中,设置一个文本标签和 3 个按钮,分别对应消息提示 Toast 的 3 种显示方式。代码如下:

```xml
1  <?xml version="1.0" encoding="utf-8"?>
2  <LinearLayout xmlns:android="http://schemas.android.com/apk/res/android"
3      android:layout_width="fill_parent"
4      android:layout_height="fill_parent"
5      android:orientation="vertical" >
6     <TextView
7         android:layout_width="fill_parent"
8         android:layout_height="wrap_content"
9         android:gravity="center"          ← 居中显示文本
10        android:text="消息提示 Tost"
11        android:textSize="24sp" />
12    <Button
13        android:id="@+id/btn1"
14        android:layout_height="wrap_content"
15        android:layout_width="fill_parent"
16        android:text="默认方式"
17        android:textSize="20sp" />
18    <Button
19        android:id="@+id/btn2"
20        android:layout_height="wrap_content"
21        android:layout_width="fill_parent"
22        android:text="自定义方式"
23        android:textSize="20sp" />
24    <Button
25        android:id="@+id/btn3"
26        android:layout_height="wrap_content"
27        android:layout_width="fill_parent"
28        android:text="带图标方式"
29        android:textSize="20sp" />
30 </LinearLayout>
```

(2) 设计控制文件 MainActivity.java 的代码如下:

```java
1  package com.ex03_15;
2  import android.app.Activity;
3  import android.os.Bundle;
4  import android.view.Gravity;
```

```
5   import android.view.View;
6   import android.view.View.OnClickListener;
7   import android.widget.Button;
8   import android.widget.ImageView;
9   import android.widget.LinearLayout;
10  import android.widget.ListView;
11  import android.widget.Toast;
12
13  public class MainActivity extends Activity
14  {
15     ListView list;
16     Button btn1,btn2,btn3;
17     @Override
18     public void onCreate(Bundle savedInstanceState)
19     {
20        super.onCreate(savedInstanceState);
21        setContentView(R.layout.activity_main);
22        btn1=(Button)findViewById(R.id.btn1);
23        btn2=(Button)findViewById(R.id.btn2);
24        btn3=(Button)findViewById(R.id.btn3);
25        btn1.setOnClickListener(new mClick());
26        btn2.setOnClickListener(new mClick());           ◄── 为按钮注册事件监听器
27        btn3.setOnClickListener(new mClick());
28     }
29
30     class mClick implements OnClickListener
31     {
32        Toast toast;
33        LinearLayout toastView;
34        ImageView imageCodeProject;
35        @Override
36        public void onClick(View v)
37        {
38           if(v==btn1)       ◄── 居中显示文本
39           {
40              Toast.makeText(getApplicationContext(),   ◄── 设置提示消息内容，可
41                      "默认Toast方式",                      用 MainActivity.this 替换
42                      Toast.LENGTH_SHORT).show();           getApplicationContext()
43           }
44           else if(v==btn2)
45           {
46              toast=Toast.makeText(getApplicationContext(),
47                      "自定义Toast的位置",
48                      Toast.LENGTH_SHORT);
49              toast.setGravity(Gravity.CENTER, 0, 0);   ◄── 自定义显示位置
```

```
50              toast.show();
51          }
52          else if(v==btn3)
53          {
54              toast=Toast.makeText(getApplicationContext(),
55                          "带图标的Toast",
56                          Toast.LENGTH_SHORT);
57              toast.setGravity(Gravity.CENTER, 0, 80);
58              toastView=(LinearLayout)toast.getView();        ← 定义视图
59              imageCodeProject=new ImageView(MainActivity.this);
60              imageCodeProject.setImageResource(R.drawable.icon); ← 获取资源中的图标
61              toastView.addView(imageCodeProject, 0);         ← 在视图中添加图标
62              toast.show();
63          }
64      }
65  }
66  }
```

程序的运行结果如图 3.21 所示。

图 3.21　消息提示 Tost 的 3 种方式

3.8　列表组件

3.8.1　列表组件 ListView 类

ListView 类是 Android 程序开发中经常用到的组件，该组件必须与适配器配合使用，由适配器提供显示样式和显示数据。

ListView 类的常用方法见表 3-16。

表 3-16 ListView 类的常用方法

常 用 方 法	说 明
ListView(Context context)	构造方法
setAdapter(ListAdapter adapter)	设置提供数组选项的适配器
addHeaderView(View v)	设置列表项目的头部
addFooterView(View v)	设置列表项目的底部
setOnItemClickListener(AdapterView.OnItemClickListener listener)	注册单击选项时执行的方法，该方法继承于父类 android.widget.AdapterView

【例 3-16】列表组件示例。

在界面设计中，设置一个文本标签和一个列表组件 ListView。

程序设计步骤：

（1）在界面布局文件中声明列表组件 ListView。

（2）在 Activity 中获得相关组件实例。

（3）通过触发列表的选项事件，调用 mClick 类的 onClick() 方法显示相应提示内容。

程序代码：

（1）设计界面布局文件 activity_main.xml。

```
1   <?xml version="1.0" encoding="utf-8"?>
2   <LinearLayout xmlns:android="http://schemas.android.com/apk/res/android"
3       android:layout_width="fill_parent"
4       android:layout_height="fill_parent"
5       android:orientation="vertical" >
6       <TextView
7           android:layout_width="fill_parent"
8           android:layout_height="wrap_content"
9           android:text="凤凰传奇"
10          android:textSize="24sp" />
11      <ListView
12          android:id="@+id/ListView01"
13          android:layout_height="wrap_content"
14          android:layout_width="fill_parent" />
15  </LinearLayout>
```

（2）设计控制文件 MainActivity.java。

```
1   package com.ex03_16;
2   import android.app.Activity;
3   import android.os.Bundle;
4   import android.view.View;
5   import android.widget.AdapterView;
```

```java
6  import android.widget.AdapterView.OnItemClickListener;
7  import android.widget.ArrayAdapter;
8  import android.widget.ListView;
9  import android.widget.TextView;
10 import android.widget.Toast;
11 public class MainActivity extends Activity
12 {
13   ListView list;
14   @Override
15   public void onCreate(Bundle savedInstanceState)
16   {
17      super.onCreate(savedInstanceState);
18      setContentView(R.layout.activity_main);
19      list= (ListView)findViewById(R.id.ListView01);    // 与界面的列表组件建立关联
20      //定义数组
21      String[] data ={
22                      "(1)荷塘月色",
23                      "(2)最炫民族风",
24                      "(3)天蓝蓝",
25                      "(4)最美天下",
26                      "(5)自由飞翔",
27                      };
28      //为 ListView 设置数组适配器 ArrayAdapter
29      list.setAdapter(new ArrayAdapter<String>(this,
30        android.R.layout.simple_list_item_1, data));      // 为列表设置适配器和监听器
31      //为 ListView 设置列表选项监听器
32      list.setOnItemClickListener(new mItemClick());
33   }
34   //定义列表选项监听器的事件
35   class mItemClick implements OnItemClickListener
36   {
37     @Override
38    public void onItemClick(AdapterView<?> arg0, View arg1, int arg2, long arg3)
39    {
40      Toast.makeText(MainActivity,"您选择的项目是: "     // 提示信息
41           +((TextView)arg1).getText(), Toast.LENGTH_SHORT).show();
42    }
43   }
44 }
```

语句说明：

（1）android.R.layout.simple_list_item_1 是一个 Android 系统内置的 ListView 布局方式。

- android.R.layout.simple_list_item_1：一行 text。
- android.R.layout.simple_list_item_2：一行 title，一行 text。
- android.R.layout.simple_list_item_single_choice：单选按钮。
- android.R.layout.simple_list_item_multiple_choice：多选按钮。

（2）OnItemClickListener 是个接口，用于监听列表组件选项的触发事件。

（3）Toast.makeText().show()显示提示消息框。

程序的运行结果如图 3.22 所示。

图 3.22　列表组件示例

3.8.2　列表组件 ListActivity 类

当整个 Activity 中只有一个 ListView 组件时，可以使用 ListActivity。其实，ListActivity 和一个只包含一个 ListView 组件的普通 Activity 没有太大区别，只是实现了一些封装而已。ListActivity 类继承于 Activity 类，默认绑定了一个 ListView 组件，并提供了一些与 ListView 处理相关的操作。

ListActivity 类常用的方法为 getListView()，该方法返回绑定的 ListView 组件。

【例 3-17】ListActivity 应用示例。

（1）设计界面布局文件 activity_main.xml，其代码如下：

```
1  <?xml version="1.0" encoding="utf-8"?>
2  <LinearLayout xmlns:android="http://schemas.android.com/apk/res/android"
3    android:layout_width="fill_parent"
4    android:layout_height="fill_parent"
5    android:orientation="vertical" >
6    <ListView
7      android:id="@+id/android:list"
8      android:layout_height="wrap_content"
```

```
9    android:layout_width="fill_parent" />
10  </LinearLayout>
```

说明：

ListActivity 布局文件中的 ListView 组件的 id 应设为"@id/android:list"。

（2）设计控制文件 MainActivity.java，其代码如下：

```
1   package com.ex03_17;
2   import android.app.ListActivity;
3   import android.os.Bundle;
4   import android.view.View;
5   import android.widget.AdapterView;
6   import android.widget.ArrayAdapter;
7   import android.widget.ListView;
8   import android.widget.TextView;
9   import android.widget.Toast;
10  import android.widget.AdapterView.OnItemClickListener;
11  public class MainActivity extends ListActivity
12  {
13   @Override
14   public void onCreate(Bundle savedInstanceState)
15   {
16    super.onCreate(savedInstanceState);
17    setContentView(R.layout.activity_main);
18   //定义数组
19    String[] data={
20     "（1）荷塘月色",
21     "（2）最炫民族风",
22     "（3）天蓝蓝",
23     "（4）最美天下",
24     "（5）自由飞翔",
25     };
26   //获取列表项
27    ListView list=getListView();
28   //设置列表项的头部
29    TextView header=new TextView(this);
30    header.setText("凤凰传奇经典歌曲");
31    header.setTextSize(24);
32    list.addHeaderView(header);
33   //设置列表项的底部
34    extView foot=new TextView(this);
35    foot.setText("请选择");
36    foot.setTextSize(24);
```

```
37    list.addFooterView(foot);
38    setListAdapter(new ArrayAdapter<String>(this,
39    android.R.layout.simple_list_item_1, data));
40    list.setOnItemClickListener(new mItemClick());
41    }
42    //定义列表选项监听器
43    class mItemClick implements OnItemClickListener
44    {
45    @Override
46    public void onItemClick(AdapterView<?> arg0, View arg1, int arg2, long arg3)
47    {
48    Toast.makeText(getApplicationContext(),
49        "您选择的项目是: "+((TextView)arg1).getText(),
50        Toast.LENGTH_SHORT).show();
51    }
52    }
53 }
```

程序的运行结果如图 3.23 所示。

图 3.23 ListActivity 应用示例

3.9 滑动抽屉组件

在日常生活中，当杂乱的物品很多时，可以把这些物品分类整理好放在不同的抽屉中，这样在使用物品时，打开抽屉，里面的东西一目了然。在 Android 系统中，也可以把多个程序放到一个应用程序的抽屉里。如图 3.24（a）所示，单击"向上"按钮（称为手柄）时，打开抽屉；如图 3.24（b）所示，单击"向下"按钮时，关闭抽屉。

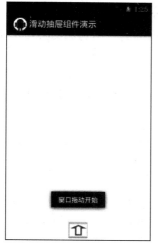　　　　　　　　

（a）单击"向上"按钮，将打开抽屉　　　　（b）单击"向下"按钮，将关闭抽屉

图 3.24　滑动抽屉示例

使用 Android 系统提供的 SlidingDraw 组件可以实现滑动抽屉的功能。先来看一下 SlidingDraw 类的重要方法和属性，其重要的 XML 属性如表 3-17 所示。

表 3-17　SlidingDraw 类重要的 XML 属性

属　性	说　明
android:allowSingleTap	设置通过手柄打开或关闭滑动抽屉
android:animateOnClick	单击手柄时，是否加入动画，默认为 true
android:handle	指定抽屉的手柄 handle
android:content	隐藏在抽屉里的内容
android:orientation	滑动抽屉内的对齐方式

SlidingDraw 类的重要方法如表 3-18 所示。

表 3-18　SlidingDraw 类的重要方法

方　法	说　明
animateOpen()	关闭时实现动画
animateOpen()	打开时实现动画
getContent()	获取内容
getHandle()	获取手柄
setOnDrawerOpenListener(SlidingDrawer.OnDrawerOpenListener onDrawerOpenListener)	打开抽屉的监听器
setOnDrawerCloseListener(SlidingDrawer.OnDrawerCloseListener onDrawerCloseListener)	关闭抽屉的监听器
setOnDrawerScrollListener(SlidingDrawer.OnDrawerScrollListener onDrawerScrollListener)	打开/关闭切换时的监听器

【例 3-18】实现图 3.24 所示的滑动抽屉 SlidingDraw 组件的应用示例。

事先准备好两个图标文件，分别命名为 up.jpg 和 down.jpg，将它们复制到 res\drawable-hdpi 目录下，以作滑动抽屉的手柄之用。

（1）设计界面布局文件 Activity_main.xml。

在 XML 文件中设置一个 SlidingDraw 组件，然后设置一个图标按钮 ImageButton 作为抽屉手柄。activity_main.xml 的代码如下：

```xml
1  <LinearLayout xmlns:android="http://schemas.android.com/apk/res/android"
2      xmlns:tools="http://schemas.android.com/tools"
3      android:id="@+id/LinearLayout1"
4      android:layout_width="match_parent"
5      android:layout_height="match_parent"
6      android:orientation="vertical" >
7      <!--设置 handle 和 content 的 id-->
8      <SlidingDrawer
9          android:layout_width="fill_parent"
10         android:layout_height="fill_parent"
11         android:handle="@+id/handle"
12         android:content="@+id/content"
13         android:orientation="vertical"
14         android:id="@+id/slidingdrawer"  >
15     <!--设置 handle,就是用一个图标按钮来处理滑动抽屉事件-->
16     <ImageButton
17         android:id="@id/handle"
18         android:layout_width="50dip"
19         android:layout_height="44dip"
20         android:src="@drawable/up" />
21     <!--设置抽屉内容,当拖动抽屉的时候就会看到-->
22     <LinearLayout
23         android:id="@id/content"
24         android:layout_width="fill_parent"
25         android:layout_height="fill_parent"
26         android:background="#66cccc"
27         android:focusable="true" >
28     </LinearLayout>
29     </SlidingDrawer>
30 </LinearLayout>
```

（2）设计控制文件 MainActivity.java。在控制程序 MainActivity.java 中，主要是实现滑动抽屉的几个监听事件。其代码如下：

```java
1  package com.example.ex03_18;
2  import android.os.Bundle;
3  import android.app.Activity;
4  import android.widget.ArrayAdapter;
5  import android.widget.ImageButton;
6  import android.widget.LinearLayout;
```

```java
7   import android.widget.ListView;
8   import android.widget.SlidingDrawer;
9   import android.widget.Toast;
10  
11  public class MainActivity extends Activity
12  {
13      SlidingDrawer mDrawer;
14      ImageButton imgBtn;
15      ListView listView;
16      LinearLayout layout;
17      String data[]=new String[]{"使命召唤","植物大战僵尸","愤怒的小鸟"};
18      @Override
19      public void onCreate(Bundle savedInstanceState)
20      {
21          super.onCreate(savedInstanceState);
22          setContentView(R.layout.activity_main);
23          layout=(LinearLayout) findViewById(R.id.content);
24          listView = new ListView(MainActivity.this);
25          listView.setAdapter(new ArrayAdapter<String>(
26              MainActivity.this,
27              android.R.layout.simple_expandable_list_item_1,
28              data));
29          layout.addView(listView);
30          imgBtn=(ImageButton)findViewById(R.id.handle);
31          mDrawer=(SlidingDrawer)findViewById(R.id.slidingdrawer);
32          mDrawer.setOnDrawerOpenListener(new mOpenListener());
33          mDrawer.setOnDrawerCloseListener(new mCloseListener());
34          mDrawer.setOnDrawerScrollListener(new mScrollListener());
35      }
36  
37      class mOpenListener implements SlidingDrawer.OnDrawerOpenListener
38      {
39        @Override
40        public void onDrawerOpened()
41        {
42            imgBtn.setImageResource(R.drawable.down);
43        }
44      }
45  
46      class mCloseListener implements SlidingDrawer.OnDrawerCloseListener
47      {
48        @Override
49        public void onDrawerClosed()
50        {
51            imgBtn.setImageResource(R.drawable.up);
52        }
53      }
```

```
54
55    class mScrollListener implements SlidingDrawer.OnDrawerScrollListener
56    {
57      @Override
58      public void onScrollEnded()
59      {
60        Toast.makeText(MainActivity.this, "结束拖动",
61                              Toast.LENGTH_SHORT).show();
62      }
63      @Override
64      public void onScrollStarted()
65      {
66        Toast.makeText(MainActivity.this, "窗口拖动开始",
67                              Toast.LENGTH_SHORT).show();
68      }
69    }
70  }
```

打开/关闭切换时触发

习 题 3

1. 编写程序，实现如图 3.25 所示的功能，即单击按钮，在文本编辑框中输入的文字内容显示到文本标签中。

图 3.25 文本编辑框中的文字内容显示到文本标签中

2. 设计一个加法计算器，如图 3.26 所示，在前两个文本编辑框中输入整数，单击"＝"按钮时，在第 3 个文本编辑框中显示这两个数之和。

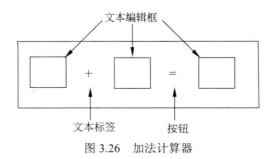

图 3.26 加法计算器

3. 设计如图 3.27 所示的用户界面布局。

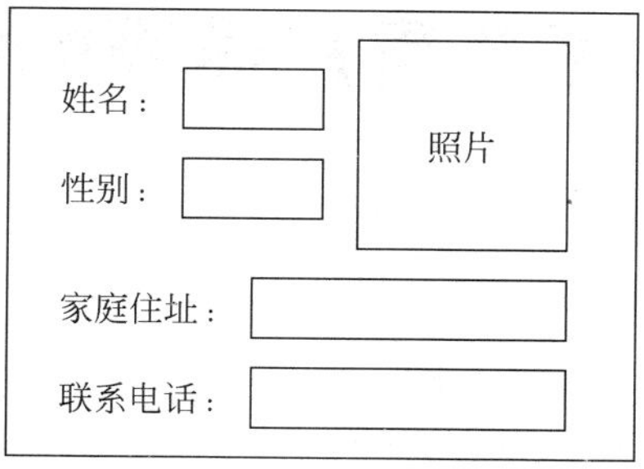

图 3.27　设计用户界面

4. 参照例 3-14，将画廊的小图片显示到屏幕的下方，在图片切换器中显示的放大图片显示在屏幕的上方。

第 4 章　多个用户界面的程序设计

4.1　页面切换与传递参数值

4.1.1　传递参数组件 Intent

　　Intent 是 Android 系统一种运行时的绑定机制，在应用程序运行时连接两个不同组件。在 Android 的应用程序中不管是页面切换，还是传递数据，或是调用外部程序，都可能要用到 Intent。Intent 负责对应用中某次操作的动作、动作涉及的数据和附加数据进行描述，Android 则根据此 Intent 的描述，负责找到对应的组件，将 Intent 传递给调用的组件，并完成组件的调用。因此，可以将 Intent 理解为不同组件之间通信的"媒介"，它专门提供组件互相调用的相关信息。

　　Intent 的属性有动作（Action）、数据（Data）、分类（Category）、类型（Type）、组件（Compent）及扩展（Extra）。其中，最常用的是 Action 属性。

　　例如：

- Intent.ACTION_MAIN 表示标识 Activity 为一个程序的开始。
- Intent.ACTION_GET_CONTENT 表示允许用户选择图片或录音等特殊类型的数据。
- Intent.ACTION_SEND 表示发送邮件的 Action 动作。
- Telephony.SMS_RECEIVED 表示接收邮件的 Action 动作。
- Intent.ACTION_ANSWER 表示处理呼入的电话。
- Intent.Action_CALL_BUTTON 表示按"拨号"键。
- Intent.Action_CALL 表示呼叫指定的电话号码。

4.1.2　Activity 页面切换

　　Activity 跳转与传递参数值主要通过 Intent 类实现。在一个 Activity 页面中启动另一个 Activity 页面的运行，是最简单的 Activity 页面切换方式。其步骤如下：

　　（1）创建一个 Intent 对象，其构造方法为：

`Intent intent = new Intent(当前Activity.this,另一个Activity.class);`

　　（2）调用 Activity 的 startActivity（intent）方法，切换到另一个 Activity 页面。

　　【例 4-1】从一个 Activity 页面启动另一个 Activity 页面示例。

　　创建名为 Ex04_01 的新项目，包名为 com.ex04_01。在本项目中，要建立

视频演示

两个页面文件及两个控制文件：第一个页面的界面布局文件为 activity_main.xml，控制文件为 MainActivity.java；第二个页面的界面布局文件为 second.xml，控制文件为 secondActivity.java。另外，还要修改配置文件 AndroidManifest.xml。

1. 设计第一个页面

（1）修改第一个页面的控制文件 MainActivity.java，代码如下：

```
1  package com.ex04_01;
2  import android.app.Activity;
3  import android.content.Intent;
4  import android.os.Bundle;
5  import android.view.View;
6  import android.view.View.OnClickListener;
7  import android.widget.Button;
8  public class MainActivity extends Activity
9  {
10    private Button btn;
11    @Override
12    public void onCreate(Bundle savedInstanceState)
13    {
14      super.onCreate(savedInstanceState);
15      setContentView(R.layout.activity_main);
16      btn=(Button)findViewById(R.id.mButton);
17      btn.setOnClickListener(new btnclock());
18    }
19    class btnclock implements OnClickListener    ← 定义一个类实现监听接口
20    {
21      public void onClick(View v)
22      {
23        Intent intent=new Intent(MainActivity.this, secondActivity.class);
24        //创建 Intent 之后,即可通过它启动新的 Activity
25        startActivity(intent);
26      }
27    }
28  }
```

（2）第一个页面的界面布局文件 activity_main.xml 的代码如下：

```
1  <?xml version="1.0" encoding="utf-8"?>
2  <LinearLayout xmlns:android="http://schemas.android.com/apk/res/android"
3    android:layout_width="fill_parent"
4    android:layout_height="fill_parent"
5    android:orientation="vertical" >
6    <TextView
7      android:id="@+id/textView1"
8      android:layout_width="fill_parent"
9      android:layout_height="wrap_content"
10     android:text="@string/hello" />
```

```
11  <Button
12    android:id="@+id/mButton"
13    android:layout_width="wrap_content"
14    android:layout_height="wrap_content"
15    android:text="@string/button"
16  />
17 </LinearLayout>
```

2. 设计第二个页面

(1) 在项目中新建第二个页面的控制文件 secondActivity.java。

首先右击包资源管理器中的 ex04_01\src\com.ex04_01 项，在弹出的快捷菜单中选择"新建"→"文件"命令，如图 4.1 所示。

图 4.1　新建一个 Java 源程序

然后在弹出的"新建文件"对话框中输入文件名 secondActivity.java，并输入代码如下：

```
1  package com.ex04_01;
2  import android.app.Activity;
3  import android.os.Bundle;
4  public class secondActivity extends Activity
5  {
6    @Override
7    public void onCreate(Bundle savedInstanceState)
8    {
9      super.onCreate(savedInstanceState);
10     setContentView(R.layout.second);    ← 启动界面布局文件 second.xml
11   }
12 }
```

(2) 新建第二个页面的界面布局文件 second.xml。

其操作步骤同前，即右击包资源管理器中的 ex04_01\res\layout 项，在弹出的快捷菜单中选择"新建"→"文件"命令，然后在"新建文件"对话框中输入文件名 second.xml，并输入代码如下：

```
1  <?xml version="1.0" encoding="utf-8"?>
2  <LinearLayout xmlns:android="http://schemas.android.com/apk/res/android"
3      android:layout_width="fill_parent"
4      android:layout_height="fill_parent"
5      android:orientation="vertical" >
6      <TextView
7          android:layout_width="fill_parent"
8          android:layout_height="wrap_content"
9          android:text="@string/second" />
10 </LinearLayout>
```

3. 修改 strings.xml 和配置文件 AndroidManifest.xml

（1）strings.xml 文件的代码如下：

```
1  <?xml version="1.0" encoding="utf-8"?>
2  <resources>
3      <string name="hello">切换页面</string>
4      <string name="app_name">Ex04_01</string>
5      <string name="second"> 这是第二个页面 </string>
6      <string name="button">切换到另一页面</string>
7  </resources>
```

（2）修改 AndroidManifest.xml 配置文件。打开项目中的 AndroidManifest.xml 文件，向其注册第二个 Activity 页面，其代码如下：

```
1  <?xml version="1.0" encoding="utf-8"?>
2  <manifest xmlns:android="http://schemas.android.com/apk/res/android"
3      package="com.ex04_01"
4      android:versionCode="1"
5      android:versionName="1.0" >
6      <uses-sdk android:minSdkVersion="15" />
7      <application
8          android:icon="@drawable/ic_launcher"
9          android:label="@string/app_name" >
10         <activity
11             android:label="@string/app_name"
12             android:name=".MainActivity" >
13             <intent-filter >
14                 <action android:name="android.intent.action.MAIN" />
15                 <category android:name="android.intent.category.LAUNCHER" />
16             </intent-filter>
17         </activity>
18         <activity
19             android:label="@string/app_name"
20             android:name=".secondActivity" />
21     </application>
22 </manifest>
```

← 新增第二个 Activity 的注册

程序的运行结果如图 4.2 所示。

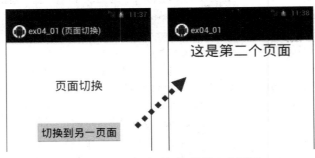

图 4.2　从一个页面切换到另一个页面

4.1.3　应用 Intent 在 Activity 页面之间传递数据

1. Bundle 类

Bundle 类是一个用于将字符串与某组件对象建立映射关系的组件。Bundle 组件与 Intent 配合使用，可在不同的 Activity 之间传递数据。Bundle 类的常用方法如下。

- putString(String key, String value)：把字符串用"键-值"形式存放到 Bundle 对象中。
- remove(String key)：移除指定 key 的值。
- getString(String key)：获取指定 key 的字符。

2. 应用 Intent 在不同的 Activity 之间传递数据

下面说明应用 Intent 与 Bundle 相配合从一个 Activity 页面传递数据到另一个 Activity 页面的方法。

1）在页面 Activity A 端

- 创建 Intent 对象和 Bundle 对象：

```
Intent intent=new Intent();
Bundle bundle=new Bundle();
```

- 为 Intent 指定切换页面，用 Bundle 存放"键-值"对数据：

```
intent.setClass(MainActivity.this, secondActivity.class);
bundle.putString("text", txt.getText().toString());
```

- 将 Bundle 对象传递给 Intent：

```
intent.putExtras(bundle);
```

2）在页面 Activity B 端

- 从 Intent 中获取 Bundle 对象：

```
bunde=this.getIntent().getExtras();
```

- 从 Bundle 对象中按"键-值"对的键名获取对应数据值：

```
String str=bunde.getString("text");
```

在不同的 Activity 页面之间传递数据的过程如图 4.3 所示。

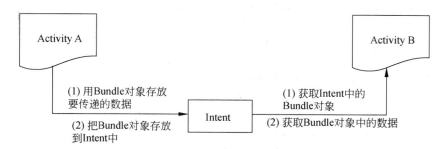

图 4.3　应用 Intent 在 Activity 页面之间传递数据

视频演示

【例 4-2】从第一个 Activity 页面传递数据到第二个 Activity 页面示例。

（1）第一个 Activity 页面的界面布局文件 activity_main.xml 的代码如下：

```
1  <?xml version="1.0" encoding="utf-8"?>
2  <LinearLayout xmlns:android="http://schemas.android.com/apk/res/android"
3      android:layout_width="fill_parent"
4      android:layout_height="fill_parent"
5      android:orientation="vertical" >
6      <TextView
7          android:layout_width="fill_parent"
8          android:layout_height="wrap_content"
9          android:text="页面切换" />
10     <Button
11         android:id="@+id/button1"
12         android:layout_width="wrap_content"
13         android:layout_height="wrap_content"
14         android:text="切换到另一页面" />
15     <EditText
16         android:id="@+id/editText1"
17         android:layout_width="match_parent"
18         android:layout_height="wrap_content" >
19         <requestFocus />       ← 获得焦点
20     </EditText>
21 </LinearLayout>
```

（2）第一个 Activity 页面的控制文件 MainActivity.java 的代码如下：

```
1  package com.ex04_02;
2  import android.app.Activity;
3  import android.content.Intent;
4  import android.os.Bundle;
5  import android.view.View;
6  import android.view.View.OnClickListener;
7  import android.widget.Button;
```

```java
8    import android.widget.EditText;
9    public class MainActivity extends Activity
10   {
11    Button btn;
12    @Override
13    public void onCreate(Bundle savedInstanceState)
14    {
15      super.onCreate(savedInstanceState);
16      setContentView(R.layout.activity_main);
17      btn = (Button)findViewById(R.id.button1);
18      btn.setOnClickListener(new btnclock());
19    }
20      //定义一个类实现监听接口
21    class btnclock implements OnClickListener
22    {
23     public void onClick(View v)
24     {
25      Intent intent=new Intent();              // 创建 Intent 对象并指定切换页面
26      intent.setClass(mainActivity.this, secondActivity.class);
27      EditText txt=(EditText)findViewById(R.id.editText1);
28      Bundle bundle=new Bundle();              // 创建 Bundle 对象存放"键-值"对数据
29      bundle.putString("text", txt.getText().toString());
30      intent.putExtras(bundle);                // 将 Bundle 对象传递给 Intent
31      startActivity(intent);                   // 启动另一个 Activity 页面
32     }
33    }
34   }
```

（3）第二个页面的界面布局文件 second.xml 的代码如下：

```xml
1  <?xml version="1.0" encoding="utf-8"?>
2  <LinearLayout xmlns:android="http://schemas.android.com/apk/res/android"
3    android:layout_width="fill_parent"
4    android:layout_height="fill_parent"
5    android:orientation="vertical" >
6    <Button
7      android:id="@+id/button2"
8      android:layout_width="wrap_content"
9      android:layout_height="wrap_content"
10     android:text="返回第一个页面" />
11   <TextView
12     android:id="@+id/TextView2"
13     android:layout_width="match_parent"
14     android:layout_height="wrap_content"
```

```
15        android:textSize="24sp"    />
16  </LinearLayout>
```

（4）第二个页面的控制文件 secondActivity.java 的代码如下：

```
1   package com.ex04_02;
2   import android.app.Activity;
3   import android.content.Intent;
4   import android.os.Bundle;
5   import android.util.Log;
6   import android.view.View;
7   import android.view.View.OnClickListener;
8   import android.widget.Button;
9   import android.widget.TextView;
10  public class secondActivity extends Activity
11  {
12    Button btn2;
13    @Override
14    public void onCreate(Bundle savedInstanceState)
15    {
16     super.onCreate(savedInstanceState);
17     setContentView(R.layout.second);
18     TextView txt2=(TextView)findViewById(R.id.TextView2);
19     Bundle bunde=this.getIntent().getExtras();      ◄── 取得 Intent 中的 Bundle 对象
20
21     String str = bunde.getString("text");           ◄── 获取 Bundle 对象中的数据
22     txt2.setText(str);
23     btn2=(Button)findViewById(R.id.button2);
24     btn2.setOnClickListener(new btnclock2() );
25    }
26    //定义返回前一页面的监听接口事件
27    class btnclock2 implements OnClickListener
28    {
29     public void onClick(View v)
30     {
31        Intent intent2=new Intent();                 ◄── 新建 Intent 对象
32        intent2.setClass(secondActivity.this, MainActivity.class);
33        startActivityForResult(intent2, 0);          ◄── 返回前一页面
34     }
35    }
36  }
```

（5）修改配置文件 AndroidManifest.xml，同例 4-1。

程序的运行结果如图 4.4 所示。

图 4.4　数据在不同 Activity 页面之间传递

4.2　菜　　单

一个菜单（Menu）由多个菜单项组成，选择一个菜单项就可以引发一个动作事件。

在 Android 系统中，菜单可以分为 3 类：选项菜单（Option Menu）、上下文菜单（Context Menu）及子菜单（Sub Menu）。下面主要介绍选项菜单和上下文菜单的设计方法，由于子菜单的设计方法基本与选项菜单相同，这里就不赘述了。

4.2.1　选项菜单

选项菜单需要通过按下设备的 Menu 键来显示。当按下设备上的 Menu 键后，在屏幕底部会弹出一个菜单，这个菜单称为选项菜单（OptionsMenu）。

1. 在 Activity 中创建菜单的方法

设计选项菜单需要用到 Activity 的 onCreateOptionMenu(Menu menu)方法，用于建立菜单并且在菜单中添加菜单项。另外，还需要用到 Activity 的 onOptionsItemSelected(MenuItem item)方法，用于响应菜单事件。Activity 实现选项菜单的方法见表 4-1。

表 4-1　Activity 实现选项菜单的方法

方　　法	说　　明
onCreateOptionMenu(Menu menu)	用于初始化菜单，menu 为 Menu 对象的实例
onPrepareOptionsMenu(Menu menu)	改变菜单状态，在菜单显示前调用
onOptionsMenuClosed(Menu menu)	菜单被关闭时调用
onOptionsItemSelected(MenuItem item)	菜单项被单击时调用，即菜单项的监听方法

2. 菜单（Menu）

设计选项菜单需要用到 Menu、MenuItem 接口。一个 Menu 对象代表一个菜单，在 Menu 对象中可以添加菜单项（MenuItem），也可以添加子菜单（SubMenu）。

菜单使用 add(int groupId, int itemId, int order, CharSequence title) 方法添加一个菜单项。add()方法中的 4 个参数依次是：

（1）组别。如果不分组就写 Menu.NONE。

（2）ID。这个参数很重要，Android 根据这个 ID 来确定不同的菜单。

（3）顺序。哪个菜单项在前面由这个参数的大小决定。

（4）文本。菜单项的显示文本。

3. 创建选项菜单的步骤

创建选项菜单的步骤如下：

（1）重写 Activity 的 onCreateOptionMenu(Menu menu)方法，当菜单第一次被打开时调用。

（2）调用 Menu 的 add()方法添加菜单项（MenuItem）。

（3）重写 Activity 的 onOptionsItemSelected(MenuItem item)方法，当菜单项被选择时响应事件。

【例 4-3】选项菜单应用示例。

设计一个选项菜单应用的示例程序，其运行结果如图 4.5 所示。

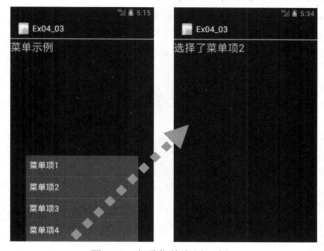

视频演示

图 4.5　选项菜单应用示例

其控制文件 MainActivity.java 的代码如下：

```
1 package com.ex04_03;
2 import android.app.Activity;
3 import android.os.Bundle;
4 import android.view.Menu;
5 import android.view.MenuItem;
6 import android.widget.TextView;
7 public class MainActivity extends Activity
8 {
9  TextView txt;
10    @Override
11    public void onCreate(Bundle savedInstanceState)
12    {
13       super.onCreate(savedInstanceState);
14       setContentView(R.layout.activity_main);
15     txt=(TextView)findViewById(R.id.TextView1);
16    }
17    @Override
```

```
18    public boolean onCreateOptionsMenu(Menu menu)
19    {
20        //调用父类方法加入系统菜单
21      super.onCreateOptionsMenu(menu);
22       //添加菜单项
23      menu.add(
24                1,           //组号
25                1,           //唯一的ID号
26                1,           //排序号
27                "菜单项1");//标题
28      menu.add( 1, 2, 2,   "菜单项2");
29      menu.add( 1, 3, 3,   "菜单项3");
30      menu.add( 1, 4, 4,   "菜单项4");
31        return true;
32    }
33     @Override
34    public boolean onOptionsItemSelected(MenuItem item)
35    {
36      String title="选择了" + item.getTitle().toString();
37      switch (item.getItemId())
38       { //响应每个菜单项(通过菜单项的ID)
39        case 1:
40              txt.setText(title);
41              break;
42        case 2:
43              txt.setText(title);
44              break;
45        case 3:
46              txt.setText(title);
47              break;
48        case 4:
49              txt.setText(title);
50              break;
51        default:
52              //对于没有处理的事件,交给父类来处理
53              return super.onOptionsItemSelected(item);
54       }
55      return true;
56    }
57  }
```

（箭头标注：添加菜单项的4个参数；文本标签显示菜单项的标题）

4.2.2 上下文菜单

Android 系统的上下文菜单类似于计算机上的右键菜单。当为一个视图注册了上下文菜单后,长按(两秒左右)这个视图对象就会弹出一个浮动菜单,即上下文菜单。任何视

图都可以注册上下文菜单，不过，最常见的是用于列表视图 ListView 的 item。

创建上下文菜单的步骤如下：

（1）重写 Activity 的 onCreateContenxtMenu()方法，调用 Menu 的 add 方法添加菜单项（MenuItem）。

（2）重写 Activity 的 onContextItemSelected()方法，响应上下文菜单菜单项的单击事件。

（3）调用 Activity 的 registerForContextMenu()方法，为视图注册上下文菜单。

【例 4-4】上下文菜单应用示例。

设计一个上下文菜单应用的示例程序，其运行结果如图 4.6 所示。

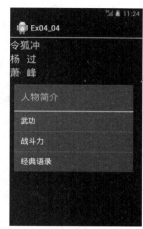

图 4.6 上下文菜单应用示例

其控制文件 MainActivity.java 的代码如下：

```
1   package com.ex04_04;
2   import android.app.Activity;
3   import android.os.Bundle;
4   import android.view.ContextMenu;
5   import android.view.ContextMenu.ContextMenuInfo;
6   import android.view.Menu;
7   import android.view.MenuItem;
8   import android.view.View;
9   import android.widget.ListView;
10  import android.widget.TextView;
11  public class MainActivity extends Activity
12  {
13    TextView txt1, txt2, txt3;
14    private static final int item1=Menu.FIRST;
15    private static final int item2=Menu.FIRST+1;
16    private static final int item3=Menu.FIRST+2;
17    String str[]={" 令狐冲", "杨 过", "萧 峰" };
18    @Override
19    public void onCreate(Bundle savedInstanceState)
20    {
21      super.onCreate(savedInstanceState);
22      setContentView(R.layout.activity_main);
23      txt1=(TextView)findViewById(R.id.textView1);
24      txt2=(TextView)findViewById(R.id.textView2);
25      txt3=(TextView)findViewById(R.id.textView3);
26      txt1.setText(str[0].toString());
27      txt2.setText(str[1].toString());
28      txt3.setText(str[2].toString());
29      registerForContextMenu(txt1);
30      registerForContextMenu(txt2);
31      registerForContextMenu(txt3);
32    }
```

← 初始化组件

```
33      //上下文菜单,本例会通过长按条目激活上下文菜单
34      @Override
35      public void onCreateContextMenu(ContextMenu menu, View view,
36      ContextMenuInfo menuInfo) {
37          menu.setHeaderTitle("人物简介");
38          //添加菜单项
39          menu.add(0, item1, 0, "武功");
40          menu.add(0, item2, 0, "战斗力");
41          menu.add(0, item3, 0, "经典语录");
42      }
43      //菜单单击响应
44      @Override
45      public boolean onContextItemSelected(MenuItem item){
46          //获取当前被选择的菜单项的信息
47          switch(item.getItemId())
48          {
49            case item1:
50                //在这里添加处理代码
51                break;
52            case item2:
53                //在这里添加处理代码
54                break;
55            case item3:
56                //在这里添加处理代码
57                break;
58          }
59          return true;
60      }
61  }
```

（选项的具体功能没有实现）

4.3 对 话 框

对话框是一个有边框、有标题栏的独立存在的容器，在应用程序中经常使用对话框组件来进行人机交互。Android 系统提供了 4 种常用对话框。

- AlertDialog：消息对话框；
- ProgressDialog：进度条对话框；
- DatePickerDialog：日期选择对话框；
- TimePickerDialog：时间选择对话框。

下面逐一介绍这些对话框的使用方法。

4.3.1 消息对话框

消息对话框（AlertDialog）是应用程序设计中最常用的对话框之一。消息对话框的内容

很丰富，使用消息对话框可以创建普通对话框、带列表的对话框以及带单选按钮和多选按钮的对话框。消息对话框的常用方法如表 4-2 所示。

表 4-2　消息对话框的常用方法

方　　法	说　　明
AlertDialog.Builder(Context)	对话框 Builder 对象的构造方法
create();	创建 AlertDialog 对象
setTitle();	设置对话框的标题
setIcon();	设置对话框的图标
setMessage();	设置对话框的提示信息
setItems();	设置对话框要显示的一个 List
setPositiveButton();	在对话框中添加 yes 按钮
setNegativeButton();	在对话框中添加 no 按钮
Show();	显示对话框
dismiss();	关闭对话框

创建消息对话框对象需要使用消息对话框的内部类 Builder。设计消息对话框的步骤如下。

（1）用 AlertDialog.Builder 类创建对话框的 Builder 对象：

```
Builder dialog=new AlertDialog.Builder(Context);
```

（2）设置对话框的标题、图标、提示信息、按钮等：

```
dialog.setTitle("普通对话框");
dialog.setIcon(R.drawable.icon1);
dialog.setMessage("一个简单的提示对话框");
dialog.setPositiveButton("确定", new okClick());
```

（3）创建并显示消息对话框的对象：

```
dialog.create();
dialog.show();
```

如果在对话框内部设置了按钮，还需要对其设置事件监听 OnClickListener。

【例 4-5】消息对话框应用示例。

在本例中设计了两种形式的对话框程序，一个是发出提示信息的普通对话框，另一个是用户登录对话框。

在用户登录对话框中，设计了用户登录的界面布局文件 long.xml，供用户输入相关验证信息。

程序的运行结果如图 4.7 所示。

（a）普通对话框　　　　　　（b）用户登录对话框

图 4.7　消息对话框

（1）界面布局文件 activity_main.xml 的代码如下：

```
1   <?xml version="1.0" encoding="utf-8"?>
2   <LinearLayout xmlns:android="http://schemas.android.com/apk/res/android"
3       android:layout_width="fill_parent"
4       android:layout_height="fill_parent"
5       android:gravity="center_horizontal"
6       android:orientation="vertical" >
7       <Button
8           android:id="@+id/button1"
9           android:layout_width="wrap_content"
10          android:layout_height="wrap_content"
11          android:text="打开普通对话框"
12          android:textSize="20sp"
13          />
14      <Button
15          android:id="@+id/button2"
16          android:layout_width="wrap_content"
17          android:layout_height="wrap_content"
18          android:text="打开输入对话框"
19          android:textSize="20sp"
20          />
21  </LinearLayout>
```

（2）用户登录对话框的界面布局文件 login.xml 的代码如下：

```
1   <?xml version="1.0" encoding="utf-8"?>
2   <LinearLayout xmlns:android="http://schemas.android.com/apk/res/android"
3       android:layout_width="fill_parent"
4       android:layout_height="fill_parent"
```

```
5       android:orientation="vertical" >
6       <TextView
7           android:id="@+id/user"
8           android:layout_width="fill_parent"
9           android:layout_height="wrap_content"
10          android:text="用户名"
11          android:textSize="18sp"/>
12      <EditText
13          android:id="@+id/userEdit"
14          android:layout_width="fill_parent"
15          android:layout_height="wrap_content"
16          android:textSize="18sp"/>
17      <TextView
18          android:id="@+id/password"
19          android:layout_width="fill_parent"
20          android:layout_height="wrap_content"
21          android:text="密码"
22          android:textSize="18sp"/>
23      <EditText
24          android:id="@+id/paswdEdit"
25          android:layout_width="fill_parent"
26          android:layout_height="wrap_content"
27          android:textSize="18sp"/>
28  </LinearLayout>
```

（3）控制文件 MainActivity.java 的代码如下：

```
1   package com.ex04_05;
2   import android.app.Activity;
3   import android.app.AlertDialog;
4   import android.app.AlertDialog.Builder;
5   import android.app.ProgressDialog;
6   import android.content.DialogInterface;
7   import android.os.Bundle;
8   import android.view.View;
9   import android.view.View.OnClickListener;
10  import android.widget.Button;
11  import android.widget.EditText;
12  import android.widget.LinearLayout;
13  import android.widget.Toast;
14
15  public class MainActivity extends Activity
16  {
17    ProgressDialog mydialog;
18     Button btn1,btn2;
19    LinearLayout login;
```

```
20    @Override
21    public void onCreate(Bundle savedInstanceState)
22    {
23       super.onCreate(savedInstanceState);
24       setContentView(R.layout.activity_main);
25       btn1=(Button)findViewById(R.id.button1);
26       btn2=(Button)findViewById(R.id.button2);
27       btn1.setOnClickListener(new mClick());
28       btn2.setOnClickListener(new mClick());
29    }
30   class mClick implements OnClickListener
31   {
32  Builder dialog=new AlertDialog.Builder(MainActivity.this);
33  @Override
34  public void onClick(View arg0)
35  {
36   if(arg0==btn1)
37   {
38      //设置对话框的标题
39      dialog.setTitle("警告");
40      //设置对话框的图标
41      dialog.setIcon(R.drawable.icon1);           ◀── 设置对话框
42      //设置对话框显示的内容
43      dialog.setMessage("本项操作可能导致信息泄露!");
44      //设置对话框的"确定"按钮
45      dialog.setPositiveButton("确定", new okClick());
46      //创建对话框
47      dialog.create();
48      //显示对话框
49      dialog.show();
50   }
51   else  if(arg0 == btn2)
52   {
53     login = (LinearLayout)getLayoutInflater()      ◀── 从另外的布局中关联组件
54             .inflate(R.layout.login, null);
55     dialog.setTitle("用户登录").setMessage("请输入用户名和密码")
56             .setView(login);
57     dialog.setPositiveButton("确定", new loginClick());
58     dialog.setNegativeButton("退出", new exitClick());
59     dialog.setIcon(R.drawable.icon2);
60     dialog.create();
61     dialog.show();
62   }
63  }
```

```
64     }
65     /*普通对话框的"确定"按钮事件*/
66     class okClick implements DialogInterface.OnClickListener
67     {
68       @Override
69       public void onClick(DialogInterface dialog, int which)
70       {
71           dialog.cancel();          ← 关闭对话框
72       }
73     }
74     /*输入对话框的"确定"按钮事件*/
75     class loginClick implements DialogInterface.OnClickListener
76     {
77       EditText txt;
78       @Override
79       public void onClick(DialogInterface dialog, int which)
80       {
81         txt=(EditText)login.findViewById(R.id.paswdEdit);   ← 关联布局文件中的组件
82         //取出输入编辑框的值与密码admin作比较
83         if((txt.getText().toString()).equals("admin"))
84             Toast.makeText(getApplicationContext(),
85                 "登录成功", Toast.LENGTH_SHORT).show();        ← 密码为admin时,显示"登录成功"
86         else
87             Toast.makeText(getApplicationContext(),
88                 "密码错误", Toast.LENGTH_SHORT).show();
89         dialog.dismiss();        ← 关闭对话框
90       }
91     }
92     /*输入对话框的"退出"按钮事件*/
93     class exitClick implements DialogInterface.OnClickListener
94     {
95       @Override
96       public void onClick(DialogInterface dialog, int which)
97       {
98           MainActivity.this.finish();    ← 单击"退出"按钮,退出MainActivity程序
99       }
100    }
101  }
```

语句说明：

在程序的第53、54行中，inflate是将组件从一个XML定义的布局中找出来。

在一个Activity中，如果直接用findViewById()，对应的是setConentView()中的layout中的组件（程序第24行中的R.layout.activity_main）。如果Activity中用到其他layout布局，例如对话框上的layout，还要设置对话框上的layout中的组件（像图片ImageView、文字

TextView）上的内容，此时必须用 inflate()先将对话框上的 layout 找出来，然后用这个 layout 对象找到它上面的组件。

4.3.2 其他几种常用对话框

1. 进度条对话框

Android 系统有一个 ProgressDialog 类，它继承于 AlertDialog，综合了进度条与对话框的特点，使用起来非常方便。ProgressDialog 类的继承关系如图 4.8 所示。

```
java.lang.Object
　└─ android.app.Dialog
　　　└─ android.app.AlertDialog
　　　　　└─ android.app.ProgressDialog
```

图 4.8 ProgressDialog 类继承于 AlertDialog

进度条对话框的常用方法见表 4-3。

表 4-3 进度条对话框的常用方法

方　　法	说　　明
getMax()	获取对话框进度的最大值
getProgress()	获取对话框当前的进度值
onStart()	开始调用对话框
setMax(int max)	设置对话框进度的最大值
setMessage(CharSequence message)	设置对话框的文本内容
setProgress(int value)	设置对话框当前的进度
show(Context context, CharSequence title, CharSequence message)	设置对话框的显示内容和方式
ProgressDialog(Context context)	对话框的构造方法

2. 日期选择对话框和时间选择对话框

日期选择对话框（DatePickerDialog）和时间选择对话框（TimePickerDialog）都继承于 AlertDialog，一般用于日期和时间的设定，它们的常用方法如表 4-4 所示。

表 4-4 日期选择对话框和时间选择对话框的常用方法

方　　法	说　　明
updateDate(int year, int monthOfYear, int dayOfMonth)	设置 DatePickerDialog 对象的当前日期
onDateChanged(DatePicker view, int year, int month, int day)	修改 DatePickerDialog 对象的日期
updateTime(int hourOfDay, int minutOfHour)	设置 TimePickerDialog 对象的时间
onTimeChanged(TimePicker view, int hourOfDay, int minute)	修改 TimePickerDialog 对象的时间

【例 4-6】 进度及日期、时间对话框示例。

```
1   package com.example.ex04_06;
2   import android.app.Activity;
3   import android.app.DatePickerDialog;
4   import android.app.ProgressDialog;
5   import android.app.TimePickerDialog;
6   import android.app.DatePickerDialog.OnDateSetListener;
7   import android.app.TimePickerDialog.OnTimeSetListener;
8   import android.os.Bundle;
9   import android.view.View;
10  import android.view.View.OnClickListener;
11  import android.widget.Button;
12  import android.widget.DatePicker;
13  import android.widget.TimePicker;
14
15  public class MainActivity extends Activity
16  {
17    Button btn1,btn2,btn3;
18    @Override
19    public void onCreate(Bundle savedInstanceState)
20    {
21        super.onCreate(savedInstanceState);
22        setContentView(R.layout.activity_main);
23        btn1=(Button)findViewById(R.id.button1);
24        btn2=(Button)findViewById(R.id.button2);
25        btn3=(Button)findViewById(R.id.button3);
26        btn1.setOnClickListener(new mClick());
27        btn2.setOnClickListener(new mClick());
28        btn3.setOnClickListener(new mClick());
29    }
30    class mClick implements OnClickListener
31    {
32      int m_year=2012;
33      int m_month=1;
34      int m_day=1;
35      int m_hour=12,  m_minute=1;
36      @Override
37      public void onClick(View v)
38      {
39       if(v==btn1)
40         {
41            ProgressDialog  d=new ProgressDialog (MainActivity.this);
42            d.setTitle("进度对话框");
43            d.setIndeterminate(true);
44            d.setMessage("程序正在 Loading...");
```

```java
45            d.setCancelable(true);
46            d.setMax(10);
47            d.show();
48        }
49        else if(v==btn2)
50        {
51            //设置日期监听器
52            OnDateSetListener dateListener = new OnDateSetListener()
53            {
54                @Override
55                public void onDateSet(DatePicker view, int year,
56                                      int monthOfYear, int dayOfMonth)
57                {
58                    m_year=year;
59                    m_month=monthOfYear;
60                    m_day=dayOfMonth;
61                }
62            };
63            //创建日期对话框对象
64            DatePickerDialog date = new DatePickerDialog(MainActivity.this,
65                    dateListener, m_year, m_month, m_day);
66            date.setTitle("日期对话框");
67            date.show();
68        }
69        else if(v==btn3)
70        {   //设置时间监听器
71            OnTimeSetListener timeListener = new OnTimeSetListener()
72            {
73                @Override
74                public void onTimeSet(TimePicker view, int hourOfDay, int minute)
75                {
76                    m_hour=hourOfDay;
77                    m_minute=minute;
78                }
79            };
80            TimePickerDialog d = new TimePickerDialog(MainActivity.this,
81                            timeListener, m_hour, m_minute, true);
82            d.setTitle("时间对话框");
83            d.show();
84        }
85    }
86  }
87 }
```

程序的运行结果如图 4.9 所示。

（a）打开进度对话框　　（b）打开日期对话框

图 4.9　对话框示例

习　题　4

1．设计一个具有两个页面的程序，第一个页面显示一张封面的图片，第二个页面显示"欢迎进入本系统"，这两个页面之间能相互切换。

2．设计一个具有 3 个选项的菜单程序，当单击每个选项时，分别跳转到 3 个不同的页面。

3．设计一个具有计算器功能的对话框程序。

第 5 章　异常处理与多线程

5.1　异　常　处　理

异常（Exception）指程序运行过程中出现的非正常现象，例如用户输入错误、需要处理的文件不存在、在网络上传输数据但网络没有连接等。由于异常情况总是可能发生，良好健壮的应用程序除了具备用户所要求的基本功能外，还应该具备预见并处理可能发生的各种异常的功能。所以，开发应用程序时要充分考虑到各种可能发生的异常情况，使程序具有较强的容错能力。通常把这种对异常情况进行处理的技术称为异常处理。

在 Android 系统中应用 Java 语言的异常处理机制进行异常处理。

1. 异常处理机制

在 Android 系统的异常处理中，引入了一些用来描述和处理异常的 Java 类，每个异常类反映一类运行错误，在类的定义中包含了该类异常的信息和对异常进行处理的方法。当程序运行过程中发生某个异常现象时，系统会产生一个与之相对应的异常类对象，并交由系统中的相应机制进行处理，以避免系统崩溃或其他对系统有害的结果发生，保证了程序运行的安全性。这就是 Android 系统的异常处理机制。

2. 异常类的定义

在 Android 系统中，按 Java 语言对异常的分类，把异常分为错误（Error）与异常（Exception）两大类。

错误（Error）通常是指程序本身存在非法的情形，这些情形常常是因为代码存在问题而引起的。而且，编程人员可以通过对程序进行更加仔细的检查，把这种错误的情形减到最小。从理论上讲，这些情形可以避免。

异常情况（Exception）表示另一种"非同寻常"的错误。这种错误通常是不可预测的。常见的异常情况包括内存不足、找不到所需的文件等。

Throwable 类派生了两个子类：Exception 和 Error。其中，Error 类描述内部错误，它由系统保留，程序不能抛出这个类型的对象，Error 类的对象不可捕获、不可以恢复，出错时系统通知用户并终止程序；而 Exception 类则供应程序使用。所有的 Android 异常类都是系统类库中 Exception 类的子类。

同其他类一样，Exception 类有自己的方法和属性。它的构造方法有两个：

```
public Exception();
public Exception(String s);
```

第二个构造方法接受字符串参数传入的信息，该信息通常是对异常所对应的错误的描述。
Exception 类从父类 Throwable 那里还继承了若干种方法，其中常用的方法有：

（1）public String toString()。toString()方法返回描述当前 Exception 类信息的字符串。

（2）public void printStackTrace()。printStackTrace()方法没有返回值，它的功能是完成打印操作，在当前的标准输出（一般是屏幕）上打印输出当前异常对象的堆栈使用轨迹，即程序先后调用执行了哪些对象，或类的哪些方法使得运行过程中产生了这个异常对象。

3. 系统定义的运行异常

Exception 类有若干个子类，每个子类代表一种特定的运行错误。有些子类是系统事先定义的，并包含在 Android 系统的 Java 类库中，称为系统定义的运行异常。

系统定义的运行异常通常对应系统运行错误。由于这种错误可能导致操作系统错误甚至是整个系统的瘫痪，所以需要定义异常类来特别处理。表 5-1 中列出了若干常见的系统定义异常。

表 5-1 Android 系统引用 Java 定义的运行异常类

系统定义的运行异常	说　明
ClassNotFoundException	找不到要装载的类，由 Class.forName 抛出
ArrayIndexOutOfBoundsException	数组下标出界
FileNotFoundException	找不到指定的文件或目录
IOException	输入/输出错误
NullPointerException	非法使用空引用
ArithmeticException	算术错误，如除数为 0
InterruptedException	一个线程被另一个线程中断
UnknownHostException	无法确定主机的 IP 地址
SecurityException	安全性错误
MalfomedURLException	URL 格式错误

由于定义了相应的异常，程序即使产生某些致命的错误，如应用空对象等，系统也会自动产生一个对应的异常对象来处理和控制这个错误，避免其蔓延或产生更大的问题。

【例 5-1】异常处理示例：当除数为 0 时，抛出异常。

```
1   package com.ex05_01;
2   import android.app.Activity;
3   import android.os.Bundle;
4   import android.app.ProgressDialog;
5   import android.widget.TextView;
6   import android.widget.Toast;
7
8   public class MainActivity extends Activity {
9       TextView txt1;
10      ProgressDialog log;
11      Toast toast;
```

```
12      public void onCreate(Bundle savedInstanceState) {
13          super.onCreate(savedInstanceState);
14          txt1=new TextView(this);
15          int x=15,y=0,z;
16          String c=" ";
17          txt1.setText("运行结果 ");
18          try{  z=x/y;  c=String.valueOf(z);  }
19          catch(Exception e)
20              {  c="x=15, y=0, z=x/y 错误，除数不能为 0! ";}
21          toast=Toast.makeText(this, c, Toast.LENGTH_LONG);   ← Toast 使用静态方法
22          toast.setText(c);
23          toast.show();
24       setContentView(txt1);
25          }
26  }
```

程序的运行结果如图 5.1 所示。

图 5.1　除数为 0 时抛出异常

5.2　多　线　程

5.2.1　线程与多线程

1. 线程概述

线程是指进程中单一顺序的执行流。设某程序的地址空间在 0x0000～0xffff，其线程 A 运行在 0x2000～0x4000，线程 B 运行在 0x4001～0x6000，线程 C 运行在 0x6001～0x8000，多个线程共同构成一个大的进程，如图 5.2 所示。

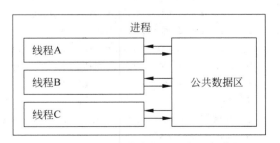

图 5.2　每个线程彼此独立，但有公共数据区

线程间的通信非常简单而有效，上下文切换非常快，它们是同一个进程中其中两部分之间所进行的切换。每个线程彼此独立执行，一个程序可以同时使用多个线程来完成不同的任务。一般用户在使用多线程时并不需要考虑底层处理的详细细节。

2. 多线程概述

多线程是指一个程序中包含有多个执行流,多线程是实现并发机制的一种有效手段。

例如,在传统的单进程环境下,用户必须等待一个任务完成后才能进行下一个任务。即使大部分时间空闲,也只能按部就班地工作。而多线程可以避免用户的等待。

又如,传统的并发服务器是基于多线程机制的,每个客户需要一个进程,而进程的数目是受操作系统限制的。基于多线程的并发服务器,每个客户一个线程,多个线程可以并发执行。

进程与多线程的区别如图 5.3 所示。

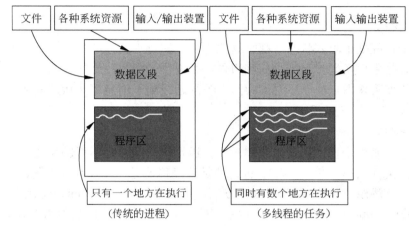

图 5.3 进程与多线程的区别

从图中可以看到,多任务状态下各进程的内部数据和状态都是完全独立的,而多线程共享一块内存空间和一组系统资源,有可能相互影响。

5.2.2 线程的生命周期

每个线程都要经历创建、就绪、运行、阻塞和死亡 5 个状态,线程从产生到消失的状态变化过程称为生命周期。线程的生命周期如图 5.4 所示。

1. 创建状态

当通过 new 命令创建了一个线程对象后,该线程对象就处于创建状态。如下面语句:

```
Thread thread1=new Thread();
```

创建状态是线程已被创建但未开始执行的一个特殊状态。此时,线程对象拥有自己的内存空间,但没有分配 CPU 资源,需通过 start()方法调度进入就绪状态等待 CPU 资源。

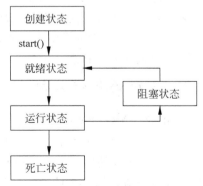

图 5.4 线程的生命周期

2. 就绪状态

处于创建状态的线程对象通过 start()方法进入就绪状态。如下面语句:

```
Thread thread1=new Thread();
Thread1.start();
```

start()方法同时调用了线程体,也就是run()方法,表示线程对象正等待CPU资源,随时可被调用执行。

处于就绪状态的线程已经被放到某一队列等待系统为其分配对CPU的控制权。至于何时真正地执行,取决于线程的优先级以及队列的当前状况。

3. 运行状态

若线程处于正在运行的状态,表示线程已经拥有了对处理器的控制权,其代码目前正在运行,除非运行过程中控制权被另一优先级更高的线程抢占,否则这一线程将一直持续到运行完毕。

4. 阻塞状态

如果一个线程处于阻塞状态,那么该线程将无法进入就绪队列。处于阻塞状态的线程通常必须由某些事件唤醒。至于是何种事件,则取决于阻塞发生的原因。例如:处于休眠中的线程必须被阻塞固定的一段时间唤醒;被挂起或处于消息等待状态的线程则必须由一外来事件唤醒。

5. 死亡状态

死亡状态(或终止状态)表示线程已退出运行状态,并且不再进入就绪队列。其原因可能是线程已执行完毕(正常结束),也可能是该线程被另一线程强行中断,即线程自然撤销或被停止。自然撤销是从线程的run()方法正常退出。即,当run()方法结束后,该线程自然撤销。调用stop()方法可以强行停止当前线程。但这个方法已在JDK2中作废,应当避免使用。如果需要线程死亡,可以进行适当的编码触发线程提前结束run()方法,使其自行消亡。

简单归纳一下,一个线程的生命周期一般经过以下几个步骤:

(1)一个线程通过new()操作实例化后,进入新生状态。

(2)通过调用start()方法进入就绪状态,一个处在就绪状态的线程将被调度执行,执行该线程相应的run()方法中的代码。

(3)通过调用线程的(或从Object类继承过来的)sleep()或wait()方法,这个线程进入阻塞状态。一个线程可能自己完成阻塞操作。

(4)当run()方法执行完毕,或者有一个例外产生,或者执行System.exit()方法,则一个线程就会进入死亡状态。

5.2.3 线程的数据通信

1. 消息(Message)

在Android的多线程中,把需要传递的数据称为消息。

由于Android的用户界面UI(User Interface)是单线程的,如果UI线程花费太多的时间做后台的事情,超过5秒钟,Android就会给出错误提示。因此,为了避免"拖累"UI,一些较费时的工作应该交给独立的后台线程去执行。但是如果后台的线程直接执行UI对象,Android会发出错误信息,所以,UI线程与后台线程需要进行消息通信。UI线程将工作分配给后台线程,后台线程执行后将相应的状态消息返回给UI线程,让UI线程对UI完成相应地更新。

Message是一个描述消息的数据结构类,Message包含很多成员变量和方法。消息的常用方法见表 5-2。

表 5-2　消息（Message）的常用方法

方　　法	说　　明
Message()	创建消息对象的构造方法
getTarget()	获取将接收此消息的 Handler 对象，此对象必须实现 Handler.handleMessage()方法
setTarget(Handler target)	设置接收此消息的 Handler 对象
sendToTarget()	向 Handler 对象发送消息
int arg1	用于当仅需要存储几个整型数据消息时
int arg2	用于当仅需要存储几个整型数据消息时
int what	用户自定义消息标识，避免各线程的消息冲突

2. 消息处理工具 Handler

Handler 是 Android 中多个线程间消息传递和定时执行任务的"工具"类。Handler 是消息的处理者，负责在多个线程之间发送 Message 和处理 Message。

Handler 类在多线程中有两个方面的应用：
- 发送消息。在不同的线程间传递数据，使用的方法为 send×××()。
- 定时执行任务。在指定的未来某时间执行某任务，使用的方法为 post×××()。

一个线程只能有一个 Handler 对象，通过该对象向所在线程发送消息。Handler 除了给其他线程发送消息外，还可以给本线程发送消息。

Handler 类的常用方法见表 5-3。

表 5-3　Handler 类的常用方法

方　　法	说　　明
Handler()	Handler 对象的构造方法
handleMessage(Message msg)	Handler 的子类必须使用该方法接收消息
sendEmptyMessage(int)	发送一个空的消息
sendMessage(Message)	发送消息，消息中可携带参数
sendMessageAtTime(Message,long)	在未来某一时间点发送消息
sendMessageDelayed(Message,long)	延时 N 毫秒发送消息
post(Runnable)	提交计划任务马上执行
postAtTime(Runnable,long)	提交计划在未来的时间点执行
postDelayed(Runnable,long)	提交计划任务延时 N 毫秒执行

应用 Handler 对象处理线程发送消息的一般形式如下。

（1）在线程的 run()方法中发送消息：

```
public void run()
{
    Message msg=new Message();
```

```
//消息标志
    msg.what=1;
//由 Handler 对象发送这个消息
    handler.sendMessage(msg);
}
```
（2）用 Handler 对象处理消息：
```
private class mHandler extends Handler
  {
      public void handleMessage(Message msg)
      {
          switch(msg.what)
          {
            case 1:  ...
            case 2:  ...
          }
      }
  }
```
其中，handleMessage(Message msg)的参数 msg 是接收多线程 run()方法中发送的 Message 对象，msg.what 为消息标志。

5.2.4 创建线程

1. 创建线程的两种方式

在 Android 中创建线程的方法与在 Java 中创建线程的方法相同，可采用两种方式创建线程：

（1）通过创建 Thread 类的子类来构造线程。Java 定义了一个直接从根类 Object 中派生的 Thread 类，所有从这个类派生的子类或间接子类均为线程。

（2）通过实现一个 Runnable 接口的类来构造线程。

注意，Android 中的线程抛弃了 Java 线程中一些不安全的做法。例如，在 Java 中终止一个 Thread 线程，可以调用 stop()、destroy()等方法实现，但在 Android 中，这些方法都不能实现，故不能直接使用。

2. 创建 Thread 子类构造线程

可以通过继承 Thread 类建立一个 Thread 类的子类并重新设计（重载）其 run()方法来构造线程。

Thread 类是用来创建一个新线程的类。使用该类的方法，可以处理线程的优先级和改变线程的状态。

要创建和执行一个线程需完成下列步骤：

（1）创建一个 Thread 类的子类；

（2）在 Thread 子类中重新定义自己的 run()方法，在这个 run()方法中包含线程要实现的操作，并通过 Handler 对象发送 Message 消息；

（3）用关键字 new 创建一个线程对象；

（4）调用 start()方法启动线程。

线程启动后当执行 run()方法完毕时，会自然进入终止状态。

【例 5-2】创建一个多线程 Thread 类的子类，每隔 1 秒钟发送一个信号给主程序，主程序进行计数。

设计思路：

（1）创建多线程 Thread 子类 mThread。在 run()方法中通过 Thread.skeeo(1000)延时 1 秒钟，实现每隔 1 秒钟发送一次 Message 消息的功能。

（2）创建 Handler 的子类 mHandler，在 handleMessage(Message nsg)方法中接收多线程发送的 Message 消息。

（3）每接收一次消息，计数器加 1，并显示出来。

新建工程 ex05_02，在界面布局文件 activity_main.xml 中安排一个"启动线程"按钮和一个"停止线程"按钮，再安排一个文本标签，用于显示计数。界面布局如图 5.5 所示。

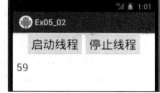

图 5.5 线程每隔 1 秒钟向主程序发送一个信号

（1）界面布局文件 activity_main.xml 的代码如下：

```
1  <LinearLayout xmlns:android="http://schemas.android.com/apk/res/android"
2      xmlns:tools="http://schemas.android.com/tools"
3      android:layout_width="fill_parent"
4      android:layout_height="fill_parent"
5      android:background="@color/white"
6      android:orientation="vertical" >
7      <LinearLayout
8          android:layout_width="fill_parent"
9          android:layout_height="wrap_content"
10         android:gravity="center_horizontal"
11         android:orientation="horizontal" >
12        <Button
13            android:id="@+id/mbutton"
14            android:layout_width="wrap_content"
15            android:layout_height="wrap_content"
16            android:text="启动线程"
17            android:textSize="20px" />
18        <Button
19            android:id="@+id/sbutton"
20            android:layout_width="wrap_content"
21            android:layout_height="wrap_content"
22            android:text="停止线程"
23            android:textSize="20px" />
24     </LinearLayout>
25     <TextView
26         android:id="@+id/txt"
27         android:layout_width="wrap_content"
```

```
28          android:layout_height="wrap_content"
29          android:layout_centerHorizontal="true"
30          android:layout_centerVertical="true"
31          android:padding="@dimen/padding_medium"
32          android:text="@string/hello_world"
33          tools:context=".MainActivity"
34          android:textSize="24dp"/>
35   </LinearLayout>
```

（2）控制文件 MainActivity.java 的代码如下：

```
1    package com.example.ex05_02;
2    import android.os.Bundle;
3    import android.os.Handler;
4    import android.os.Message;
5    import android.app.Activity;
6    import android.view.View;
7    import android.view.View.OnClickListener;
8    import android.widget.Button;
9    import android.widget.TextView;
10
11   public class MainActivity extends Activity
12   {
13       private boolean STOP=true;         ←定义线程是否停止的标志位
14       private int count=0;               ←定义按秒计时的计数器
15       private mHandler handler;
16       private mThread thread;
17       private Button mButton, sButton;
18       private TextView mTextView;
19       @Override
20       public void onCreate(Bundle savedInstanceState)
21       {
22         super.onCreate(savedInstanceState);
23         setContentView(R.layout.activity_main);
24         handler=new mHandler();
25         thread=new mThread();
26         mTextView=(TextView)findViewById(R.id.txt);
27         mButton=(Button)findViewById(R.id.mbutton);
28         mButton.setOnClickListener(new mClick());
29         sButton=(Button)findViewById(R.id.sbutton);
30         sButton.setOnClickListener(new mClick());
31   }
32       //定义监听按钮的事件,启动线程或停止线程
33       class mClick implements OnClickListener
34       {
35         @Override
36         public void onClick(View arg0)
```

```
37      {
38          if(arg0==mButton)           ┐
39          {                           │
40              //设置标志位             │── 处理"启动线程"按钮事件
41              STOP = false;           │
42              //开启新的线程           │
43              thread.start();         │
44          }                           ┘
45          else if(arg0==sButton)      ┐
46          {                           │── 处理"停止线程"按钮事件
47              STOP = true;    ← 设置标志位
48          }                           ┘
49      }
50  }
51  //定义 Handler 的子类接收和处理线程发送来的消息
52  private class mHandler extends Handler
53  {
54      public void handleMessage(Message msg)
55      {
56          switch(msg.arg1)    ← 以消息标志为条件
57          {
58              case 1:     ← 消息标志为1时，执行本复合语句
59              {
60                  count++;    ← 秒数增加
61                  mTextView.setText(Integer.toString(count));   ← 数值转换字符串
62                  break;
63              }
64          }
65      }
66  }
67  //定义 Thread 子类,实现每隔1秒钟发送一次消息的功能
68  private class mThread extends Thread
69  {
70      @Override
71      public void run()   ← 线程启动时执行这个函数
72      {
73          while(!STOP)    ← 一直循环,直到标志位为"真"
74          {
75              try{
76                      Thread.sleep(1000);     ← 延时1秒
77                  } catch(InterruptedException e){
78                      e.printStackTrace();
79                  }
80              Message msg = new Message();
81              msg.arg1=1;     ← 消息标志
```

```
82                handler.sendMessage(msg);    ← 发送消息
83            }
84        }
85    }
86 }
```

3. 实现 Runnable 接口构造线程

Runnable 接口是在程序中使用线程的另一种方法。在许多情况下，一个类已经继承了父类，因而这样的类不能再继承 Thread。Runnable 接口为一个类提供了一种手段，无须扩展 Thread 类就可以执行一个新的线程或者被一个新的线程控制。这就是通过建立一个实现 Runnable 接口的对象，并以它作为线程的目标对象来构造线程。它打破了单一继承方式的限制。

在 Java 语言的代码中，Runnable 接口只包含一个抽象方法，其定义如下：

```
public interface Runnable
{
    public abstract void run();
}
```

因此，一个类实现 Runnable 接口时需要实现多线程的 run()方法。

为了实现 Runnable 对象的线程，可使用下列方法来生成 Thread 对象：

```
Thread(Runnable 对象名);
Thread(Runnable 对象名,String 线程名);
```

【例 5-3】 应用多线程设计一个小球移动的程序。

设计一个绘制小球图形的 ballView，并由一个线程来控制小球运动。由于 ballView 要继承 View，因此，要实现多线程就需要使用 Runnable 接口。关于图形的绘制，在第 6 章有详细讲解，这里不作具体说明。

（1）控制文件 MainActivity.java 的代码如下：

```
1   package com.example.ex05_03;
2   import android.app.Activity;
3   import android.os.Bundle;
4   import android.os.Handler;
5   import android.os.Message;
6   import android.view.View;
7   import android.view.View.OnClickListener;
8   import android.widget.Button;
9
10  public class MainActivity extends Activity
11  {
12      int i=80, j=10, step;          ← i、j 为小球坐标位置；step 为移动的步长值
13      ballView view;
14      Button btn;
15      Handler handler;
16      Thread thread;
17      boolean STOP=true;             ← 线程是否停止标志
18      public void onCreate(Bundle savedInstanceState)
```

```java
19  {
20      super.onCreate(savedInstanceState);
21      setContentView(R.layout.main);
22      view=(ballView)findViewById(R.id.view1);
23      btn=(Button)findViewById(R.id.btn1);
24      btn.setOnClickListener(new mClick());
25      handler=new mHandler();
26      thread=new mThread();
27      view.setXY(i, j);
28  }
29  class mClick implements  OnClickListener
30  {
31   @Override
32   public void onClick(View arg0)
33   {
34       STOP=false;              ← 设置线程中循环的标志位
35       thread.start();          ← 开启新线程
36   }
37  }
38  private class mHandler extends Handler
39  {
40    public void handleMessage(Message msg)
41    {
42       switch(msg.what)         ← 以消息标志为条件
43       {
44         case 1:                ← 消息标志为1时，执行下面的复合语句
45         {
46            step=step+5;
47            j=j+step;
48            if(j>220)  STOP=true;   ← 设置停止线程中循环的标志位
49            break;
50         }
51        }
52       view.setXY(i, j);        ┐
53       view.invalidate();       ┘ 更新小球的坐标位置
54     }
55   }
56  private class mThread extends Thread
57  {
58    @Override
59    public void run()           ← 线程启动时执行run()函数
60    {
61        while(!STOP)            ← 一直循环，直到标志位为"真"
62        {
63           try
```

```
64              {
65                  Thread.sleep(500);        ← 延时 0.5 秒
66              }
67              catch(InterruptedException e)
68              {
69                  e.printStackTrace();
70              }
71              Message msg=new Message();
72              msg.what=1;    ← 消息标志       ← 发送消息
73              handler.sendMessage(msg);
74          }
75       }
76    }
77 }
```

（2）绘制图形文件 ballView.java 的代码如下：

```
1  package com.example.ex05_03;
2  import android.content.Context;
3  import android.graphics.Canvas;
4  import android.graphics.Color;
5  import android.graphics.Paint;
6  import android.util.AttributeSet;
7  import android.view.View;
8
9  public class ballView extends View
10 {
11   int x, y;
12   public ballView(Context context, AttributeSet attrs)
13   {
14     super(context, attrs);
15   }
16   void setXY(int _x, int _y)
17   {
18   x=_x;
19   y=_y;
20   }
21   protected void onDraw(Canvas canvas)
22   {
23     super.onDraw(canvas);
24     canvas.drawColor(Color.CYAN);
25     Paint paint = new Paint();
26     paint.setColor(Color.BLACK);
27     paint.setAntiAlias(true);              ← 绘制小球
28     canvas.drawCircle(x, y, 15, paint);
29     paint.setColor(Color.WHITE);
30     canvas.drawCircle(x-6, y-6, 3, paint);
31   }
32 }
```

（3）用户界面设计。

在用户界面程序中，除设置一个按钮组件之外，还设置了自定义的 ballView 组件。设置自定义的组件时，要注意添加包路径 com.example.ex05_03.ballView。其代码如下：

```
1  <?xml version="1.0" encoding="utf-8"?>
2  <LinearLayout xmlns:android="http://schemas.android.com/apk/res/android"
3      android:layout_width="fill_parent"
4      android:layout_height="fill_parent"
5      android:orientation="vertical" >
6      <Button
7          android:id="@+id/btn1"
8          android:layout_width="80dp"
9          android:layout_height="wrap_content"
10         android:text="开始"   />
11     <com.example.ex05_03.ballView
12         android:id="@+id/view1"
13         android:layout_width="fill_parent"
14         android:layout_height="fill_parent" />
15 </LinearLayout>
```

（11～14 行）定义自定义组件 ballView

程序的运行结果如图 5.6 所示。该示例说明了处理异步更新用户界面的方法。

图 5.6　由线程控制小球运动

习　题　5

设计两个独立线程，分别计数，如图 5.7 所示。

图 5.7　设计两个独立运行的线程

第6章　图形与多媒体处理

6.1　绘制几何图形

6.1.1　几何图形绘制类

在 Android 系统中绘制几何图形，需要用到一些绘图工具，这些绘图工具都在 android.graphics 包中。下面介绍这些绘图工具的常用方法和属性。

1. 画布（Canvas）

画布是 Android 绘制几何图形的主要工具，其常用方法见表 6-1。

表 6-1　画布（Canvas）的常用方法

方　　法	功　　能
Canvas()	创建一个空的画布，可以使用 setBitmap() 方法来设置绘制具体的画布
Canvas(Bitmap bitmap)	以 bitmap 对象创建一个画布，将内容都绘制在 bitmap 上，bitmap 不得为 null
drawColor()	设置 Canvas 的背景颜色
setBitmap()	设置具体画布
clipRect()	设置显示区域，即设置裁剪区
rotate()	旋转画布
skew()	设置偏移量
drawLine(float x1, float y1, float x2, float y2)	画从点 ($x1, y1$) 到点 ($x2, y2$) 的直线
drawCircle(float x, float y, float radius, Paint paint)	以 (x, y) 为圆心，以 radius 为半径画圆
drawRect (float x1, float y1, float x2, float y2, Paint paint)	画从左上角 ($x1, y1$) 到右下角 ($x2, y2$) 的矩形
drawText(String text, float x, float y ,Paint paint)	写文字
drawPath(Path path, Paint paint)	从一点到另一点的连接路径线段
drawBitmap(Bitmap bitmap, float left, float top, Paint paint)	将位图绘制到（left, top）位置

2. 画笔（Paint）

画笔用来描述所绘制图形的颜色和风格，如线条宽度、颜色等信息。其常用方法见表 6-2。

表 6-2　画笔（Paint）的常用方法

方　　法	功　　能
paint()	构造方法，创建一个辅助画笔对象
setColor(int color)	设置颜色
setStrokeWidth(float width)	设置画笔宽度
setTextSize(float textSize)	设置文字尺寸
setAlpha(int a)	设置透明度 alpha 值
setAntiAlias(boolean b)	除去边缘锯齿，取 true 值
paint.setStyle(Paint.Style style)	设置图形为空心（Paint.Style.STROKE）或实心（Paint.Style.FILL）

3．点到点的连线路径（Path）

当绘制由一些线段组成的图形（如三角形、四边形等）时，需要用 Path 类来描述线段路径。其常用方法见表 6-3。

表 6-3　连线路径（Path）的常用方法

方　　法	功　　能
lineTo(float x, float y)	从当前点到指定点画连线
moveTo(float x, float y)	移动到指定点
close()	关闭绘制连线路径

6.1.2　几何图形的绘制过程

在 Android 中绘制几何图形的一般过程为：
（1）创建一个 View 的子类，并重写 View 类的 onDraw()方法；
（2）在 View 的子类视图中使用画布对象 Canvas 绘制各种图形；
（3）使用 invalidate()方法刷新画面。

【例 6-1】绘制几何图形示例。

视频演示

本例继承自 android.view.View 的 TestView 类，重写 View 类的 onDraw()方法，在 onDraw()方法中运用 Paint 对象（绘笔）的不同设置值，在画布上绘制图形，分别绘制了矩形、圆形、三角形和文字。

其代码如下：

```
1  package com.ex06_01;
2  import android.app.Activity;
3  import android.content.Context;
4  import android.graphics.Canvas;
5  import android.graphics.Color;
6  import android.graphics.Paint;
7  import android.graphics.Path;
8  import android.os.Bundle;
```

```
9   import android.view.View;
10  public class MainActivity extends Activity
11  {
12    @Override
13    public void onCreate(Bundle savedInstanceState)
14    {
15        super.onCreate(savedInstanceState);
16        TestView tView=new TestView(this);
17        setContentView(tView);
18    }
19    private class TestView extends View
20    {
21      public TestView(Context context)
22      {
23          super(context);
24      }
25      /*重写onDraw()*/
26      @Override
27      protected void onDraw(Canvas canvas)
28      {
29        /*设置背景为青色*/
30        canvas.drawColor(Color.CYAN);
31        Paint paint=new Paint();
32        /*设置画笔宽度*/
33        paint.setStrokeWidth(3);
34        /*设置画空心图形*/
35        paint.setStyle(Paint.Style.STROKE);
36        /*去锯齿*/
37        paint.setAntiAlias(true);
38        /*画空心矩形（正方形）*/
39        canvas.drawRect(10,10,70,70,paint);
40        /*设置画实心图形*/
41        paint.setStyle(Paint.Style.FILL);
42        /*画实心矩形（正方形）*/
43        canvas.drawRect(100,10,170,70,paint);
44        /*设置画笔颜色为蓝色*/
45        paint.setColor(Color.BLUE);
46        /*画圆心为（100，120）、半径为30的实心圆*/
47        canvas.drawCircle(100,120,30,paint);
48        /*在上面的实心圆上画一个小白点*/
49        paint.setColor(Color.WHITE);
50        canvas.drawCircle(91,111,6,paint);
51        /*设置画笔颜色为红色*/
52        paint.setColor(Color.RED);
53        /*画三角形*/
54        Path path=new Path();
```

```
55      path.moveTo(100, 170);
56      path.lineTo(70, 230);
57      path.lineTo(130,230);
58      path.close();
59      canvas.drawPath(path,paint);
60      /*文字*/
61      paint.setTextSize(28);
62      paint.setColor(Color.BLUE);
63       canvas.drawText(getResources().getString(R.string. hello_world),
64                  30,270,paint);
65      }
66    }
67 }
```

程序的运行结果如图 6.1 所示。

【例 6-2】绘制一个可以在任意指定位置显示的小球。

设计思路：Android 系统应用程序的设计模式采用 MVC 模式，即把应用程序分为表现层（View）、控制层（Control）和业务模型层（Model）。在本示例中，按照这种模式，图形界面布局为表现层，Activity 控制程序为控制层，实现几何作图的绘制过程属于业务模型层。在业务模型层，将圆心坐标设为 (x,y)，则圆的位置随控制层任意输入的坐标值改变。

程序代码：

（1）表现层的图形界面布局程序 activity_main.xml 的代码如下：

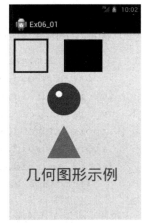

图 6.1　绘制几何图形示例

```
1  <LinearLayout xmlns:android="http://schemas.android.com/apk/res/android"
2    android:layout_width="fill_parent"
3    android:layout_height="fill_parent"
4    android:orientation="vertical">
5    <LinearLayout
6      android:layout_width="wrap_content"
7      android:layout_height="wrap_content">
8      <TextView
9         android:id="@+id/ textView1 "
10        android:layout_width="wrap_content "
11        android:layout_height="wrap_content"
12        android:text="输入位置: "
13        />
14      <EditText
15        android:id="@+id/ editText1 "
16        android:layout_width="120dp"
17        android:layout_height="wrap_content"
18        />
```

视频演示

```
19    <Button
20        android:id="@+id/button1"
21        android:layout_width="wrap_content "
22        android:layout_height="wrap_content"
23        android:text="确定"
24        />
25  </LinearLayout>
26  <com.example.ex06_02.TestView
27        android:id="@+id/testView1 "
28        android:layout_width="match_parent "
29        android:layout_height="match_parent "
30        />
31  </LinearLayout>
```

在界面布局中设置绘制图形的视图组件，导入自定义组件时，要带包名

（2）控制层的主控程序 MainActivity.java 的代码如下：

```
1   package com.example.ex06_02;
2   import android.os.Bundle;
3   import android.view.View;
4   import android.view.View.OnClickListener;
5   import android.widget.Button;
6   import android.widget.EditText;
7   import android.app.Activity;
8   public class MainActivity extends Activity
9   {
10    int x1=150,y1=50;
11    TestView testView;
12    Button btn;
13    EditText edit_y;
14    @Override
15    public void onCreate(Bundle savedInstanceState)
16    {
17        super.onCreate(savedInstanceState);
18        setContentView(R.layout.activity_main);
19        testView=(TestView)findViewById(R.id.testView1);
20        testView.setXY(x1, y1);
21        btn=(Button)findViewById(R.id.button1);
22        edit_y=(EditText)findViewById(R.id.editText1);
23        btn.setOnClickListener(new mClick());
24    }
25    class mClick implements OnClickListener
26    {
27    @Override
28    public void onClick(View arg0)
29    {
30      y1=Integer.parseInt(edit_y.getText().toString());
```

第 20 行：设置表现层图形的坐标位置

第 30 行：将字符串转换为整型

```
31         testView.setXY(x1, y1);          ◀─── 在新坐标位置绘制图形并刷新视图
32         testView.invalidate();
33     }
34 }
35 }
```

语句说明：

程序第 30 行中的方法 Integer.parseInt(String)为将字符串 String 转换为整型数据。

（3）业务模型层的绘制小球程序 TestView.java 的代码如下：

```
1  package com.example.ex06_02;
2  import android.util.AttributeSet;
3  import android.view.View;
4  import android.content.Context;
5  import android.graphics.Canvas;
6  import android.graphics.Color;
7  import android.graphics.Paint;
8
9  public class TestView extends View         ◀─── 继承于 View 的绘制图形类
10 {
11    int x, y;
12    public TestView(Context context, AttributeSet attrs)    ◀─── 在 XML 文件中使用自定义组件时必须使用 AttributeSet 接口对象做参数
13    {
14        super(context, attrs);
15    }
16    void setXY(int _x, int _y)              ◀─── 传递由控制层设置的坐标值
17    {
18        x=_x;
19        y=_y;
20    }
21    @Override
22    protected void onDraw(Canvas canvas)
23    {
24        super.onDraw(canvas);
25        /*设置背景为青色*/
26        canvas.drawColor(Color.CYAN);
27        Paint paint=new Paint();
28        /*去锯齿*/
29        paint.setAntiAlias(true);
30        /*设置 paint 的颜色*/
31        paint.setColor(Color.BLACK);
32        /*画一个实心圆*/
33        canvas.drawCircle(x, y, 15, paint);
34        /*画一个实心圆上的小白点*/
35        paint.setColor(Color.WHITE);
```

```
36            canvas.drawCircle(x-6, y-6, 3, paint);
37       }
38  }
```

程序的运行结果如图 6.2 所示。

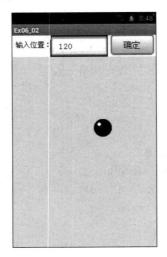

图 6.2 在任意指定位置显示小球

6.2 触摸屏事件处理

智能移动设备的触摸屏事件（模拟器中为鼠标事件）分为简单触摸屏事件和手势识别事件。下面分别介绍这些事件的处理方法。

6.2.1 简单触摸屏事件

简单触摸屏事件指在触摸屏上按下、抬起、滑动的事件（模拟器中为鼠标事件）。在 Android 系统中，通过 OnTouchListener 监听接口来处理屏幕事件，当在 View 的范围内进行按下、抬起或滑动等动作时都会触发该事件。

在设计简单触摸屏事件程序时，要实现 android.view.View.OnTouchListener 接口，并重写该接口的监听方法 onTouch(View v, MotionEvent event)。

在监听方法 onTouch(View v, MotionEvent event)中，参数 v 为事件源对象，参数 event 为事件对象，事件对象为下列常数之一。

- MotionEvent.ACTION_DOWN：按下；
- MotionEvent.ACTION_UP：抬起；
- MotionEvent.ACTION_MOVE：移动。

【例 6-3】设计一个在屏幕上移动小球的程序。

设计一个继承于 Android.view.View 的图形绘制视图 TestView，在该视图中绘制一个小球。然后设计一个实现 OnTouchListener 监听接口的类，重写该接口的监听方法 onTouch (View v, MotionEvent event)。该方法监听并

视频演示

获取触摸屏幕的坐标位置,并把坐标值传递给图形绘制类 TestView,由 TestView 在该位置重绘小球。

(1) 控制文件 MainActivity.java 的代码如下:

```
1   package com.ex06_03;
2   import android.app.Activity;
3   import android.content.Context;
4   import android.graphics.Canvas;
5   import android.graphics.Color;
6   import android.graphics.Paint;
7   import android.os.Bundle;
8   import android.util.Log;
9   import android.view.MotionEvent;
10  import android.view.View;
11  import android.view.View.OnTouchListener;
12  public class MainActivity extends Activity
13  {
14      int x1=150, y1=50;          ← 定义小球的初始坐标
15      TestView testView;
16      @Override
17      public void onCreate(Bundle savedInstanceState)
18      {
19          super.onCreate(savedInstanceState);
20          testView=new TestView(this);
21          testView.setOnTouchListener(new mOnTouch());
22          testView.getXY(x1, y1);
23          setContentView(testView);
24      }
25      private class mOnTouch implements OnTouchListener   ← 定义触摸屏事件
26      {
27          public boolean onTouch(View v, MotionEvent event)
28          {
29              if (event.getAction()==MotionEvent.ACTION_MOVE)
30              {
31                  x1=(int)event.getX();              ← 获取坐标位置
32                  y1=(int)event.getY();
33                  testView.getXY(x1, y1);            ← 按新坐标绘图
34                  setContentView(testView);
35              }
36              if (event.getAction()==MotionEvent.ACTION_DOWN)
37              {
38                  x1=(int)event.getX();              ← 获取坐标位置
39                  y1=(int)event.getY();
40                  testView.getXY(x1, y1);            ← 按新坐标绘图
41                  setContentView(testView);
42              }
```

```
43        return true;
44     }
45  }
46 }
```

（2）图形绘制类 TestView 的代码如下：

```
1  package com.ex06_03;
2  import android.content.Context;
3  import android.graphics.Canvas;
4  import android.graphics.Color;
5  import android.graphics.Paint;
6  import android.view.View;
7
8   private class TestView extends View        ← 绘制图形视图
9   {
10     int x,y;
11     public TestView(Context context)
12     {
13        super(context);
14     }
15     void getXY(int _x, int _y)              ← 由触摸屏事件传递小球的坐标位置
16     {
17        x=_x;
18        y=_y;
19     }
20     /*重写 onDraw()*/
21     @Override
22     protected void onDraw(Canvas canvas)
23     {
24        super.onDraw(canvas);
25        /*设置背景为青色*/
26        canvas.drawColor(Color.CYAN);
27        Paint paint=new Paint();
28        /*去锯齿*/
29        paint.setAntiAlias(true);
30        /*设置 paint 的颜色*/
31        paint.setColor(Color.BLACK);
32        /*画一个实心圆*/
33        canvas.drawCircle(x, y, 15, paint);
34        /*画一个实心圆上的小白点*/         ← 绘制小球
35        paint.setColor(Color.WHITE);
36        canvas.drawCircle(x-6, y-6, 3, paint);
37     }
38   }
39 }
```

程序运行结果如图 6.3 所示，用手指（或鼠标）在屏幕上滑动时，小球将随手指移动。当单击屏幕时，小球将被移到单击位置。

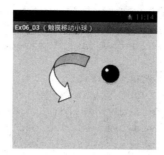

图 6.3　用手指（或鼠标）在屏幕上滑动，小球随之移动

【例 6-4】 设计一个能在图片上涂鸦的程序。

（1）界面布局文件的代码如下：

视频演示

```
1  <?xml version="1.0" encoding="utf-8"?>
2  <LinearLayout xmlns:android="http://schemas.android.com/apk/res/android"
3      android:layout_width="fill_parent"
4      android:layout_height="fill_parent"
5      android:orientation="vertical" >
6      <com.ex06_04.HandWrite
7          android:id="@+id/handwriteview"
8          android:layout_width="fill_parent"
9          android:layout_height="380dp" />
10     <LinearLayout
11         android:layout_width="fill_parent"
12         android:layout_height="fill_parent"
13         android:orientation="horizontal"
14         android:gravity="center_horizontal" >
15         <Button
16             android:id="@+id/clear"
17             android:layout_width="200dp"
18             android:layout_height="wrap_content"
19             android:text="清屏" />
20     </LinearLayout>
21 </LinearLayout>
```

导入自定义 view，注意要带包

（2）控制文件 MainActivity.java 的代码如下：

```
1  package com.ex06_04;
2  import android.app.Activity;
3  import android.os.Bundle;
4  import android.view.View;
5  import android.view.View.OnClickListener;
6  import android.widget.Button;
```

```
7   public class MainActivity extends Activity
8   {
9     private HandWrite handWrite=null;
10    private Button clear=null;
11    @Override
12    public void onCreate(Bundle savedInstanceState)
13    {
14      super.onCreate(savedInstanceState);
15      setContentView(R.layout.main);
16      handWrite=(HandWrite)findViewById(R.id.handwriteview);   ← 关联 View 组件
17      clear=(Button)findViewById(R.id.clear);
18      clear.setOnClickListener(new mClick());
19    }
20    private class mClick implements OnClickListener
21    {
22      public void onClick(View v)
23      {
24        handWrite.clear();   ← 清屏
25      }
26    }
27  }
```

（3）记录在屏幕上滑动的轨迹，实现在图片上涂鸦的功能，代码如下：

```
1   package com.ex06_04;
2   import android.content.Context;
3   import android.graphics.*;
4   import android.graphics.Paint.Style;
5   import android.util.AttributeSet;
6   import android.view.MotionEvent;
7   import android.view.View;
8   public class HandWrite extends View   ← 自定义 View 组件 HandWrite
9   {
10      Paint paint=null;                  ← 定义画笔
11      Bitmap originalBitmap=null;        ← 存放原始图像
12      Bitmap new1_Bitmap=null;           ← 存放从原始图像复制的位图图像
13      Bitmap new2_Bitmap=null;           ← 存放处理后的图像
14      float startX=0,startY=0;           ← 画线的起点坐标
15      float clickX=0,clickY=0;           ← 画线的终点坐标
16      boolean isMove=true;               ← 设置是否画线的标记
17      boolean isClear=false;             ← 设置是否清除涂鸦的标记
18      int color=Color.GREEN;             ← 设置画笔的颜色（绿色）
19      float strokeWidth=2.0f;            ← 设置画笔的宽度
20      public HandWrite(Context context, AttributeSet attrs)
21      {
22          super(context, attrs);
```

```java
23          originalBitmap=BitmapFactory                    // 从资源中获取原始图像
24              .decodeResource(getResources(), R.drawable.cy)
                .Copy(Bitmap.Config.ARGB_8888,true);
25          new1_Bitmap=Bitmap.createBitmap(originalBitmap);  // 建立原始图像的位图
26      }
27     public void clear(){
28          isClear=true;
29          new2_Bitmap=Bitmap.createBitmap(originalBitmap);  // 清除涂鸦
30          invalidate();
31      }
32     public void setstyle(float strokeWidth){
33          this.strokeWidth=strokeWidth;
34      }
35     @Override
36     protected void onDraw(Canvas canvas)
37     {
38          super.onDraw(canvas);                             // 显示绘图
39          canvas.drawBitmap(HandWriting(new1_Bitmap), 0, 0,null);
40      }
41     public Bitmap HandWriting(Bitmap o_Bitmap)             // 记录绘制图形
42     {
43          Canvas canvas=null;                               // 定义画布
44          if(isClear)
45           {
46              canvas=new Canvas(new2_Bitmap);               // 创建绘制新图形的画布
47           }
48          else{
49              canvas=new Canvas(o_Bitmap);                  // 创建绘制原图形的画布
50           }
51          paint=new Paint();
52          paint.setStyle(Style.STROKE);
53          paint.setAntiAlias(true);                         // 定义画笔
54          paint.setColor(color);
55          paint.setStrokeWidth(strokeWidth);
56     if(isMove)
57     {
58       canvas.drawLine(startX, startY, clickX, clickY, paint);  // 在画布上画线条
59      }
60     startX=clickX;
61     startY=clickY;
62     if(isClear)
63     {
64         return new2_Bitmap;                                // 返回新绘制的图像
65      }
66     return o_Bitmap;                                       // 若清屏,则返回原图像
```

```
67      }
68      @Override
69      public boolean onTouchEvent(MotionEvent event)      ← 定义触摸屏事件
70      {
71        clickX=event.getX();
72        clickY=event.getY();                              ← 获取触摸坐标位置
73        if(event.getAction() == MotionEvent.ACTION_DOWN)  ← 按下屏幕时无绘图
74        {
75          isMove=false;
76          invalidate();
77          return true;
78        }
79        else if(event.getAction() == MotionEvent.ACTION_MOVE)
80        {
81          isMove=true;                                    ← 记录在屏幕上滑动的轨迹
82          invalidate();
83          return true;
84        }
85        return super.onTouchEvent(event);
86    }
87    }
```

程序运行结果如图 6.4 所示，当用手指（或鼠标）在屏幕上滑动时，记录下滑动的轨迹，在图片上涂鸦。单击"清屏"按钮，则清除图片上的痕迹。

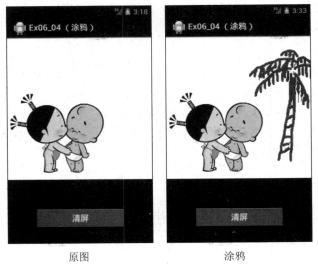

原图　　　　　　　　涂鸦

图 6.4　在图片上涂鸦

6.2.2　手势识别事件

所谓手势识别，就是识别手指（或鼠标）在屏幕上滑动时的轨迹。在 Android 系统中，

android.gesture 是用于创建、识别和保存触摸屏手势功能的包。android.gesture 包的主要类及接口如表 6-4 所示。

表 6-4　android.gesture 包的主要类及接口

类 及 接 口	功　　能
Gesture	触摸屏的手势类
GestureOverlayView	可输入手势的视图
Prediction	手势的预显示类
GestureStroke	记录触摸屏上手势动作的开始与结束类
OnGestureListener	手势动作的监听接口
OnGesturePerformedListener	可输入手势视图 GestureOverlayView 的监听接口

在实现 OnGesturePerformedListener 接口时，需要覆盖其方法：

`onGesturePerformed(GestureOverlayView overlay, Gesture gesture)`

【例 6-5】设计一个手写字体识别程序。

要编写一个手写字体识别程序，必须先建立一个存放手写字体的数据库。在手机模拟器中已经预装了创建手写字体数据库的应用程序 Gestures Builder，其图标如图 6.5 所示。

创建手势库如图 6.6 所示。由手势创建的手写字体将被保存到 sdcard\gestures 中，将文件 gestures 复制到 res\raw 下，就可以在应用程序中使用这些手势了。

视频演示

图 6.5　Gestures Builder 的图标

图 6.6　创建手势库

（1）设计界面布局文件 activity_main.xml。在界面布局文件 activity_main.xml 中设置 android.gesture.GestureOverlayView 组件。其代码如下：

```
1  <?xml version="1.0" encoding="utf-8"?>
2  <LinearLayout xmlns:android="http://schemas.android.com/apk/res/android"
3      android:layout_width="fill_parent"
```

```
4    android:layout_height="fill_parent"
5    android:orientation="vertical" >
6    <TextView
7      android:id="@+id/textView1"
8      android:layout_width="fill_parent"
9      android:layout_height="wrap_content"
10     android:text="@string/text1"
11     android:textSize="24sp"/>
12   <!-- 绘制手势的 GestureOverlayView   -->
13   <android.gesture.GestureOverlayView    ◄── 设置 View 组件，带包名
14     android:id="@+id/gestures1"
15     android:layout_width="fill_parent"
16     android:layout_height="fill_parent"
17     android:gestureStrokeType="multiple"
18     android:eventsInterceptionEnabled="false"
19     android:orientation="vertical"/>
20 </LinearLayout>
```

（2）控制文件 MainActivity.java 的代码如下：

```
1  package com.ex06_05;
2  import java.util.ArrayList;
3  import android.app.Activity;
4  import android.gesture.Gesture;
5  import android.gesture.GestureLibraries;
6  import android.gesture.GestureLibrary;
7  import android.gesture.GestureOverlayView;
8  import android.gesture.Prediction;
9  import android.gesture.GestureOverlayView.OnGesturePerformedListener;
10 import android.os.Bundle;
11 import android.widget.TextView;
12 import android.widget.Toast;
13 public class MainActivity extends Activity
14 implements OnGesturePerformedListener
15 {
16   GestureLibrary mLibrary;   ◄── 定义手势库对象
17   GestureOverlayView gesturesView;   ◄── 定义手势视图对象作画板之用
18   TextView txt;
19   @Override
20   public void onCreate(Bundle savedInstanceState)
21   {
22     super.onCreate(savedInstanceState);
23     setContentView(R.layout.main);
24     gesturesView=(GestureOverlayView)findViewById(R.id.gestures);
25     gesturesView.addOnGesturePerformedListener(this);   ◄── 注册手势识别的监听器
26     txt=(TextView)findViewById(R.id.textView1);
```

```
27    mLibrary=GestureLibraries.fromRawResource(this,
28                        R.raw.gestures);
29    if(!mLibrary.load())
30    {
31      finish();
32    }
33  }
34  /*根据在GestureOverlayView上画的手势来识别是否匹配手势库中的手势*/
35  @Override
36  public void onGesturePerformed(GestureOverlayView overlay, Gesture gesture)
37  {
38    ArrayList predictions=mLibrary.recognize(gesture);
39    if(predictions.size()>0)
40    {
41      Prediction prediction=(Prediction)predictions.get(0);
42      if(prediction.score > 1.0)
43      {
44        Toast.makeText(this,prediction.name,Toast.LENGTH_SHORT).show();
45        txt.append(prediction.name);
46      }
47    }
48  }
49 }
```

注释：加载手势库；从手势库中获取手势数据；检索到匹配的手势

程序的运行结果如图 6.7 所示。

图 6.7　手写字体识别

6.3　音频播放

6.3.1　多媒体处理包

Android 系统提供了针对常见多媒体格式的 API，使用户可以非常方便地操作图片、音频、视频等多媒体文件，也可以操纵 Android 终端的录音、摄像设备。这些多媒体处理 API 均位于 android.media 包中。android.media 包中的主要类见表 6-5。

表 6-5　android.media 包中的主要类

类名或接口名	说　明
MediaPlayer	支持流媒体，用于播放音频和视频
MediaRecorder	用于录制音频和视频
Ringtone	用于播放可用作铃声和提示音的短声音片段
AudioManager	负责控制音量
AudioRecord	用于记录从音频输入设备产生的数据
JetPlayer	用于存储 JET 内容的回放和控制
RingtoneManager	用于访问响铃、通知和其他类型的声音
Ringtone	快速播放响铃、通知或其他相同类型的声音
SoundPool	用于管理和播放应用程序的音频资源

6.3.2　媒体处理播放器

1. MediaPlayer 类的常用方法

MediaPlayer 是 Android 系统多媒体 android.media 包中的类，MediaPlayer 类主要用于控制音频文件、视频文件或流媒体的播放。MediaPlayer 类的常用方法见表 6-6。

表 6-6　MediaPlayer 类的常用方法

方　法	说　明
create()	创建多媒体播放器
getCurrentPosition()	获得当前播放位置
getDuration()	获得播放文件的时间
getVideoHeight()	播放视频高度
getVideoWidth()	播放视频宽度
isLooping()	是否循环播放
isPlaying()	是否正在播放
pause()	暂停
prepare()	准备播放文件，进行同步处理
prepareAsync()	准备播放文件，进行异步处理
release()	释放 MediaPlayer 对象
reset()	重置 MediaPlayer 对象
seekTo()	指定播放文件的播放位置
setDataSource()	设置多媒体数据来源
setVolume()	设置音量
setOnCompletionListener()	监听播放文件播放完毕
start()	开始播放
stop()	停止播放

2. MediaPlayer 对象的生命周期

通常把一个对象从创建、使用到释放该对象的过程称为该对象的生命周期，把 MediaPlayer 对象的创建、初始化、同步处理、开始播放、播放结束的运行过程称为 MediaPlayer 的生命周期，MediaPlayer 对象的生命周期如图 6.8 所示。

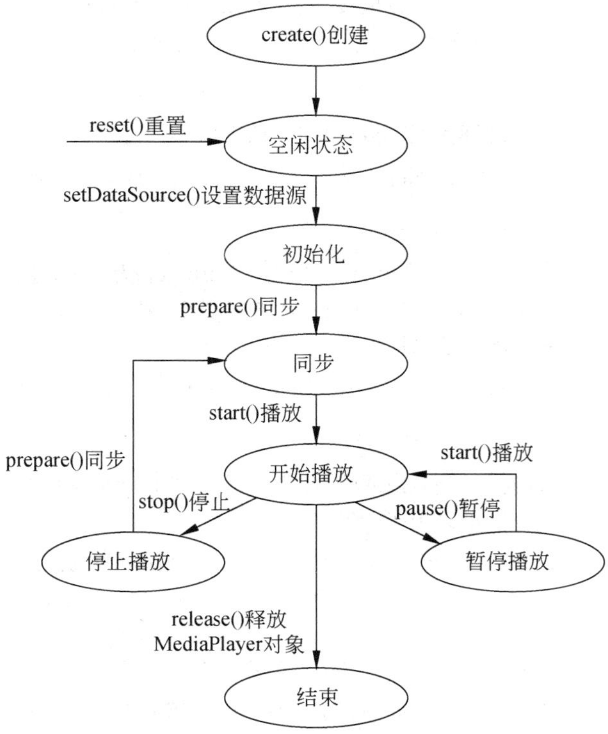

从图 6.8 可以看出，当一个 MediaPlayer 对象创建或调用了 reset() 方法后，它处于 Idle（空闲）状态。当调用了 release() 方法后，它处于释放（结束）状态。这两种状态之间是 MediaPlayer 对象的生命周期。一个 MediaPlayer 对象处于空闲状态时是不能进行播放工作的，必须经过初始化、同步阶段之后才能进行播放操作。

6.3.3 播放音频文件

通过媒体处理器 MediaPlayer 提供的方法不仅可以播放存在 SD 卡上的音乐文件，还能播放资源中的音乐文件。这二者在设计方法上稍有不同。

图 6.8 MediaPlayer 对象的生命周期

下面按两种情况来说明应用 MediaPlayer 对象播放音频文件的步骤。

1. 构建 MediaPlayer 对象

（1）使用 new 的方式创建 MediaPlayer 对象。对于播放 SD 卡上的音乐文件需要使用 new 方式来创建 MediaPlayer 对象，例如：

```
MediaPlayer mplayer=new MediaPlayer();
```

（2）使用 create() 方法创建 MediaPlayer 对象。对于播放资源中的音乐需要使用 create() 方法来创建 MediaPlayer 对象，例如：

```
MediaPlayer mplayer=MediaPlayer.create(this, R.raw.test);
```

其中，R.raw.test 为资源中的音频数据源，test 为音乐文件名称，注意不要带扩展名。

由于 create() 方法中已经封装了初始化及同步的方法，故使用 create() 方法创建的 MediaPlayer 对象不需要再进行初始化及同步工作。

2. 设置播放文件

MediaPlayer 播放的文件主要包括 3 个来源。

（1）存储在 SD 卡或其他文件路径下的媒体文件。对于存储在 SD 卡或其他文件路径下的媒体文件，需要调用 setDataSource() 方法，例如：

```
mplayer.setDataSource("/sdcard/test.mp3");
```

（2）在编写应用程序时事先存放在 res 资源中的音乐文件。播放事先存放在资源目录

res\raw 中的音乐文件，需要在使用 create()方法创建 MediaPlayer 对象时，指定资源路径和文件名称（不要带扩展名）。由于 create()方法的源代码中已经封装了调用 setDataSource()方法，因此，不必重复使用 setDataSource()方法。

（3）网络上的媒体文件。播放网络上的音乐文件，需要调用 setDataSource()方法，例如：

```
mplayer.setDataSource("http://www.citynorth.cn/music/confucius.mp3");
```

3. 对播放器进行同步控制

使用 prepare()方法设置对播放器的同步控制，例如：

```
mplayer.prepare();
```

如果 MediaPlayer 对象是由 create()方法创建的，由于 create()方法的代码中已经封装了调用 prepare()方法，因此可省略此步骤。

4. 播放音频文件

start()是真正启动音频文件播放的方法，例如：

```
mplayer.start();
```

如要暂停播放或停止播放，则调用 pause()和 stop()方法。

5. 释放占用资源

音频文件播放结束应该调用 release()释放播放器占用的系统资源。

如果要重新播放音频文件，需要调用 reset()返回到空闲状态，再从第 2 步开始重复其他各步骤。

【例 6-6】设计一个音乐播放器。

在本例中，将分别播放存放在项目资源中的音乐文件和 SD 卡中的音频文件，因此，需要事先将准备好的音频文件保存到指定路径下。

（1）将测试的音频文件 mtest1.mp3 复制到新建项目的 res\raw 目录下。

（2）将音频文件 mtest2.mp3 复制到 SD 卡中（在模拟器中使用 SD 卡，可以在 Eclipse 集成环境中选择 DDMS 调试工具，单击"向设备导入文件"按钮，将音频文件复制到模拟器的 mnt\sdcard\Music 目录下，如图 6.9 所示）。

视频演示

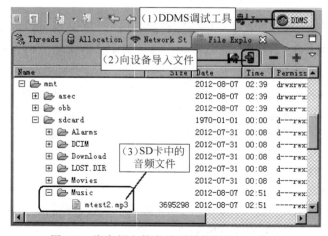

图 6.9　将音频文件存放到模拟器的 SD 卡中

（1）设计布局文件 main.xml。在用户界面布局中，设置了 3 个带图标的按钮，分别表示播放、暂停、停止，还设置了两个选项按钮，用于选择播放的文件。布局文件的代码如下：

```
1  <?xml version="1.0" encoding="utf-8"?>
2  <AbsoluteLayout
3      xmlns:android="http://schemas.android.com/apk/res/android"
4      android:orientation="vertical"
5      android:layout_width="fill_parent"
6      android:layout_height="fill_parent">
7      <TextView
8          android:id="@+id/text1"
9          android:layout_width="fill_parent"
10         android:layout_height="wrap_content"
11         android:layout_x="5px"
12         android:layout_y="10px"
13         android:text="@string/hello"
14         android:textSize="20sp"/>
15     <ImageButton               ← 带图标的按钮，需要设置图标的路径
16         android:id="@+id/Stop"
17         android:layout_height="wrap_content"
18         android:layout_width="wrap_content"
19         android:layout_x="30px"
20         android:layout_y="100px"
21         android:src="@drawable/music_stop"  />   ← 设置图标的路径和文件名称
22     <ImageButton               ← 带图标的按钮
23         android:id="@+id/Start"
24         android:layout_height="wrap_content"
25         android:layout_width="wrap_content"
26         android:layout_x="90px"
27         android:layout_y="100px"
28         android:src="@drawable/music_play"  />   ← 设置图标的路径和文件名称
29     <ImageButton               ← 带图标的按钮
30         android:id="@+id/Pause"
31         android:layout_height="wrap_content"
32         android:layout_width="wrap_content"
33         android:layout_x="150px"
34         android:layout_y="100px"
35         android:src="@drawable/music_pause"  />  ← 设置图标的路径和文件名称
36
37     <CheckBox
38         android:id="@+id/check1"
39         android:layout_width="fill_parent"
40         android:layout_height="wrap_content"
41         android:layout_x="10px"
42         android:layout_y="180px"
43         android:textSize="20sp"
44         android:text="@string/one" />
45     <CheckBox
```

```
46          android:id="@+id/check2"
47          android:layout_width="fill_parent"
48          android:layout_height="wrap_content"
49          android:layout_x="10px"
50          android:layout_y="210px"
51          android:textSize="20sp"
52          android:text="@string/two" />
53 </AbsoluteLayout>
```

（2）设计控制文件，其代码如下：

```
1  package com.ex06_06;
2  import java.io.IOException;
3  import android.app.Activity;
4  import android.media.MediaPlayer;
5  import android.os.Bundle;
6  import android.util.Log;
7  import android.view.View;
8  import android.view.View.OnClickListener;
9  import android.widget.CheckBox;
10 import android.widget.ImageButton;
11 import android.widget.TextView;
12 public class MainActivity extends Activity
13 {
14     CheckBox ch1,ch2;              ← 选项按钮
15     TextView txt;
16     ImageButton mStopButton, mStartButton, mPauseButton;  ← 播放控制按钮
17     MediaPlayer  mMediaPlayer;     ← MediaPlayer 对象
18     String sdcard_file;
19     sdcard_file=new String("/sdcard/music/mtest2.mp3");   ← SD 卡文件
20     int res_file=R.raw.mtest1;     ← 设置资源文件 mtest1
21       @Override
22     public void onCreate(Bundle savedInstanceState)
23     {
24        super.onCreate(savedInstanceState);
25        setContentView(R.layout.main);
26        /*构建 MediaPlayer 对象*/
27        mMediaPlayer=new MediaPlayer();
28        ch1=(CheckBox)findViewById(R.id.check1);
29        ch2=(CheckBox)findViewById(R.id.check2);
30        txt=(TextView)findViewById(R.id.text1);          ← 组件的初始化
31        mStopButton=(ImageButton)findViewById(R.id.Stop);
32        mStartButton=(ImageButton)findViewById(R.id.Start);
33        mPauseButton=(ImageButton)findViewById(R.id.Pause);
34        setRegist();   //动态获取本地存储卡（SD卡）读写权限
35        mStopButton.setOnClickListener(new mStopClick());
36        mStartButton.setOnClickListener(new mStartClick());
37        mPauseButton.setOnClickListener(new mPauseClick());
38     }
```

```
   //播放SD卡或其他路径的音乐文件
39 private void playMusic(String path)    ◀── 参数path为文件路径
40 {
41   try
42     {
       /*重置MediaPlayer对象,使之处于空闲状态*/
43     mMediaPlayer.reset();
       /*设置要播放文件的路径*/
44     mMediaPlayer.setDataSource(path);
       /*准备播放*/
45     mMediaPlayer.prepare();
       /*开始播放*/
46     mMediaPlayer.start();
47     }catch (IOException e){    }
48   }
   /*停止按钮事件*/
49   class mStopClick implements OnClickListener
50     {
51     @Override
52     public void onClick(View v)
53       {
         /*是否正在播放*/
54       if (mMediaPlayer.isPlaying())
55         {
           //重置MediaPlayer到初始状态
56         mMediaPlayer.reset();
57         mMediaPlayer.release();    ◀── 释放占用的资源
58         }
59       }
60     }
   /*播放按钮事件*/
61 class mStartClick implements OnClickListener
62 {
63   @Override
64   public void onClick(View v)
65     {
66     String str="";
67     if(ch1.isChecked())    ◀── 选择播放系统资源音乐
68       {
69       str=str+"\n"+ch1.getText();    ◀── 提示信息
70       try {
71         mMediaPlayer=MediaPlayer.create(MainActivity.this, res_file);
72         mMediaPlayer.start();    ◀── 开始播放系统资源中的音乐
73         } catch (Exception e) {Log.i("ch1", "res err ..."); }
74       }
75     if(ch2.isChecked())    ◀── 选择播放SD卡中的音频文件
76       {
77       str=str+"\n"+ch2.getText();    ◀── 提示信息
```

```java
78      try{
79         mMediaPlayer=new MediaPlayer();
80         mMediaPlayer.setDataSource(sdcard_file);  // 设置音乐源
81         playMusic(sdcard_file);  // 调用第38行的播放方法
82      } catch(Exception e){ Log.i("ch2", "sdcard err ... "); }
83      }
84      txt.setText(str);  // 显示提示信息
85   }
86 }
   /*暂停按钮事件*/
87 class mPauseClick implements OnClickListener
88 {
89    @Override
90    public void onClick(View v)
91    {
92       if(mMediaPlayer.isPlaying())
93       {                                          // 第一次按暂停键
          /*暂停*/
94          mMediaPlayer.pause();
95       }
96       else
97       {
          /*开始播放*/                              // 重复按暂停键
98          mMediaPlayer.start();
99       }
100   }
101 }
    //动态获取本地存储卡(SD卡)读写权限
102 private void setRegist() {
       /*
        * 动态获取权限
        * Android 6.0 新特性，一些保护权限，如，文件读写除了要在
        * AndroidManifest 中声明权限，还要使用如下代码动态获取
        */
103    if (Build.VERSION.SDK_INT >= 23) {  //大于23是指Android 6.0 以后版本
104    int REQUEST_CODE_CONTACT = 101;
105    final int REQUEST_EXTERNAL_STORAGE = 1;
106         String[] PERMISSIONS_STORAGE = {
107              Manifest.permission.READ_EXTERNAL_STORAGE,
108              Manifest.permission.WRITE_EXTERNAL_STORAGE
109    };
       //验证是否许可权限
110    for (String str : PERMISSIONS_STORAGE) {
111    if (this.checkSelfPermission(str) !=
112                           PackageManager.PERMISSION_GRANTED){
       //申请权限
113    this.requestPermissions(PERMISSIONS_STORAGE,
```

```
114                                    REQUEST_CODE_CONTACT);
115            return;
116                    }
117                }
118        }
119    }
```

程序的运行结果如图 6.10 所示。

图 6.10　音频播放示例

6.4　视　频　播　放

在 Android 系统中，设计播放视频的应用程序有两种方式，一种方式是应用媒体播放器 MediaPlayer 组件播放视频，另一种方式是应用视频视图 VideoView 组件播放视频。下面分别介绍这两种设计方式。

6.4.1　应用媒体播放器播放视频

媒体播放器 MediaPlayer 不仅可以播放音频文件，还可以播放格式为.3gp 的视频文件。与播放音频不同之处为，用于视频播放的播放承载体必须是实现了表面视图处理接口（surfaceHolder）的视图组件，即需要使用 SurfaceView 组件来显示播放的视频图像。

【例 6-7】应用媒体播放器 MediaPlayer 设计一个视频播放器。

（1）设计界面布局文件 activity_main.xml。事先准备视频文件 sample.3gp，并将其复制到模拟器的 SD 卡的 sdcard/zsm 目录下。然后在用户界面布局中，设置一个 SurfaceView 组件，用于显示视频图像。再设置一个按钮，单击该按钮，开始播放视频文件。界面布局文件的代码如下：

视频演示

```xml
1   <?xml version="1.0" encoding="utf-8"?>
2   <LinearLayout xmlns:android="http://schemas.android.com/apk/res/android"
3       android:layout_width="fill_parent"
4       android:layout_height="fill_parent"
5       android:orientation="vertical" >
6       <TextView
7           android:id="@+id/TextView01"
8           android:layout_width="wrap_content"
9           android:layout_height="wrap_content"
10          android:layout_gravity="center_horizontal"
11          android:text="媒体播放器"
12          android:textSize="24sp" />
13      <SurfaceView                                     ← 用于显示视频图像
14          android:id="@+id/surfaceView1"
15          android:layout_width="240dp"
16          android:layout_height="320dp"
17          android:layout_gravity="center" />
18      <Button
19          android:id="@+id/play1"
20          android:layout_width="80dp"
21          android:layout_height="40dp"
22          android:text="播放"
23          android:textSize="18sp"   />
24  </LinearLayout>
```

（2）设计控制文件 MainActivity.java 的代码如下：

```java
1   package com.ex06_07;
2   
3   import android.media.AudioManager;
4   import android.media.MediaPlayer;
5   import android.os.Bundle;
6   import android.util.Log;
7   import android.view.SurfaceHolder;
8   import android.view.SurfaceView;
9   import android.view.View;
10  import android.view.View.OnClickListener;
11  import android.widget.Button;
12  import android.app.Activity;
13  
14  public class MainActivity extends Activity
15  {
16      MediaPlayer mMediaPlayer;
17      SurfaceView mSurfaceView;
18      Button playBtn;
19      String path;
20      SurfaceHolder sh;            ← 表面视图处理接口对象
21  
```

```
22    @Override
23    public void onCreate(Bundle savedInstanceState)
24    {
25        super.onCreate(savedInstanceState);
26        setContentView(R.layout.Activity_main);
27        mSurfaceView=(SurfaceView)findViewById(R.id.surfaceView1);
28        playBtn=(Button)findViewById(R.id.play1);
29        path="/sdcard/zsm/sample.3gp";          ← 设定视频文件路径
30        mMediaPlayer=new  MediaPlayer();
31        playBtn.setOnClickListener(new mClick());
32    }
33
34    class mClick implements OnClickListener
35    {
36        @Override
37        public void onClick(View v)
38        {
39          try {
40              mMediaPlayer.reset();
41              //为播放器对象设置用于显示视频内容、代表屏幕描绘的控制器
42              mMediaPlayer.setAudioStreamType(AudioManager.STREAM_MUSIC);
43              mMediaPlayer.setDataSource(path); ← 设置数据源
44              sh=mSurfaceView.getHolder();      ← 创建表面视图处理接口对象
45              mMediaPlayer.setDisplay(sh);
46              mMediaPlayer.prepare();           ← MediaPlayer 对象的同步
47              mMediaPlayer.start();
48          }catch (Exception e){ Log.i("MediaPlay err", "MediaPlay err");}
49        }
50    }
51 }
```

程序的运行结果如图 6.11 所示。

图 6.11 应用媒体播放器 MediaPlayer 设计的视频播放器

6.4.2 应用视频视图播放视频

在 Android 系统中，经常使用 android.widget 包中的视频视图类 VideoView 播放视频文件。VideoView 类可以从不同的来源（例如资源文件或内容提供器）读取图像，计算和维护视频的画面尺寸，以使其适用于任何布局管理器，并提供一些诸如缩放、着色之类的显示选项。VideoView 类的常用方法见表 6-7。

表 6-7 VideoView 类的常用方法

方 法	说 明
VideoView(Context context)	创建一个默认属性的 VideoView 实例
boolean canPause()	判断是否能够暂停播放视频
int getBufferPercentage()	获得缓冲区的百分比
int getCurrentPosition()	获得当前的位置
int getDuration()	获得所播放视频的总时间
boolean isPlaying()	判断是否正在播放视频
boolean onTouchEvent (MotionEvent ev)	应用该方法来处理触屏事件
seekTo (int msec)	设置播放位置
setMediaController(MediaController controller)	设置媒体控制器
setOnCompletionListener(MediaPlayer.OnCompletionListener l)	注册在媒体文件播放完毕时调用的回调函数
setOnPreparedListener(MediaPlayer.OnPreparedListener l)	注册在媒体文件加载完毕可以播放时调用的回调函数
setVideoPath(String path)	设置视频文件的路径
setVideoURI(Uri uri)	设置视频文件的统一资源标识符
start()	开始播放视频文件
stopPlayback()	停止回放视频文件

【例 6-8】应用视频视图 VideoView 组件设计一个视频播放器。

（1）创建用户界面。新建工程 ex06_08，修改 res\layout\ activity_main.xml 界面布局文件，在其中添加一个显示视图和一个按钮。完整的界面布局文件 activity_main.xml 的代码如下：

```
1  <?xml version="1.0" encoding="utf-8"?>
2  <LinearLayout xmlns:android="http://schemas.android.com/apk/res/android"
3      android:layout_width="fill_parent"
4      android:layout_height="fill_parent"
5      android:orientation="vertical" >
6      <TextView
7          android:layout_width="fill_parent"
8          android:layout_height="wrap_content"
9          android:textSize="24sp"
10         android:text="@string/hello" />
```

视频演示

```
11    <VideoView
12        android:id="@+id/video"
13        android:layout_width="320dp"
14        android:layout_height="240dp" />
15    <Button
16        android:id="@+id/playButton"
17        android:layout_width="wrap_content"
18        android:layout_height="wrap_content"
19        android:text="@string/playButton"
20        android:textSize="20sp"    />
21 </LinearLayout>
```

（2）控制文件 MainActivity.java 的代码如下：

```
1 package com.ex06_08;
2 import android.app.Activity;
3 import android.os.Bundle;
4 import android.view.View;
5 import android.view.View.OnClickListener;
6 import android.widget.Button;
7 import android.widget.MediaController;
8 import android.widget.VideoView;
9 public class MainActivity extends Activity
10 {
11    private VideoView mVideoView;
12    private Button playBtn;
13    MediaController mMediaController;
14    @Override
15    public void onCreate(Bundle savedInstanceState)
16    {
17        super.onCreate(savedInstanceState);
18        setContentView(R.layout.Activity_main);
19        mVideoView=new VideoView(this);
20        mVideoView=(VideoView)findViewById(R.id.video);
21        mMediaController=new MediaController(this);
22        playBtn=(Button)findViewById(R.id.playButton);
23        playBtn.setOnClickListener(new mClick());
24    }
25   class mClick implements OnClickListener
26   {
27        @Override
28        public void onClick(View v)
29        {
30            String path="/sdcard/test.3gp";
31            mVideoView.setVideoPath(path);
32            mMediaController.setMediaPlayer(mVideoView);    ← 设置媒体控制器
33            mVideoView.setMediaController(mMediaController);
```

```
34            mVideoView.start();
35        }
36    }
37 }
```

程序的运行结果如图 6.12 所示。

图 6.12　应用视频视图设计的视频播放器

6.5　录音与拍照

6.5.1　用于录音、录像的 MediaRecorder 类

应用 android.media 包中的 MediaRecorder 类，可以录制音频和视频。下面详细介绍 MediaRecorder 类的使用方法。

1. MediaRecorder 类的常用方法

MediaRecorder 类的常用方法见表 6-8。

表 6-8　MediaRecorder 类的常用方法

方　　法	说　　明
MediaRecorder()	创建录制媒体对象
setAudioSource(int audio_source)	设置音频源
setAudioEncoder(int audio_encoder)	设置音频编码格式
setVideoSource(int video_source)	设置视频源
setVideoEncoder(int video_encoder)	设置视频编码格式
setVideoFrameRate(int rate)	设置视频帧速率
setVideoSize(int width, int height)	设置视频录制画面大小
setOutputFormat(int output_format)	设置输出格式

续表

方法	说明
setOutputFile(path)	设置输出文件路径
prepare()	准备录制
start()	开始录制
stop()	停止录制
reset()	重置
release()	释放播放器的有关资源

2. MediaRecorder 对象的数据采集源

（1）使用音频输入设备（麦克风）进行录音时，录音接口支持的音频源类型有：
- DEFAULT：系统音频源。
- MIC：麦克风。

（2）使用摄像设备进行视频录制时，摄像机接口所支持的视频源类型有：
- CAMERA：照相机视频输入。
- DEFAULT：平台默认。

3. MediaRecorder 对象的编码方式

（1）录音机接口支持的音频编码方式有：
- AMR_NB：AMR 窄带。
- DEFAULT：默认编码。

（2）录像机接口支持的编码方式有：
- H263：H.263 编码。
- H264：H.264 编码。
- MPEG_4_SP：MPEG4 编码。

4. MediaRecorder 对象的输出格式

- MPEG-4：MPEG4 格式。
- RAW_AMR：原始 AMR 格式文件。
- THREE_GPP：3gp 格式。

6.5.2 录音示例

应用 MediaRecorder 进行录音，其主要步骤如下：

1. 创建录音对象

```
MediaRecorder mRecorder=new MediaRecorder();
```

2. 设置录音对象

- 设置音频源：

```
mRecorder.setAudioSource(MediaRecorder.AudioSource.MIC);
```

- 设置输出格式：

```
mRecorder.setOutputFormat(MediaRecorder.OutputFormat.THREE_GPP);
```

- 设置编码格式：

```
mRecorder.setAudioEncoder(MediaRecorder.AudioEncoder.AMR_NB);
```

- 设置输出文件路径：

```
mRecorder.setOutputFile(path);
```

3. 准备录制

```
mRecorder.prepare();
```

4. 开始录制

```
mRecorder.start();
```

5. 结束录制

- 停止录制：

```
mRecorder.stop();
```

- 重置：

```
mRecorder.reset();
```

- 释放录音占用的有关资源：

```
mRecorder.release();
```

【例6-9】设计一个简易录音机。

（1）设计界面布局文件。在界面布局文件中，设置两个按钮，一个按钮用于录音，另一个按钮用于停止录音。

（2）设计控制文件 MainActivity.java 的代码如下：

```
1   package com.example.ex06_09;
2   import android.media.MediaRecorder;
3   import android.os.Bundle;
4   import android.app.Activity;
5   import android.view.View;
6   import android.view.View.OnClickListener;
7   import android.widget.Button;
8
9   public class MainActivity extends Activity
10  {
11      MediaRecorder mRecorder;
12      Button startBtn, stopBtn;
13      String path;
14      @Override
15      public void onCreate(Bundle savedInstanceState)
16      {
17          super.onCreate(savedInstanceState);
18          setContentView(R.layout.activity_main);
```

视频演示

```java
19        path="/sdcard/zsm/audio.amr";    ← 录音文件的文件名及 SD 卡的路径
20        startBtn=(Button)findViewById(R.id.button1);
21        stopBtn=(Button)findViewById(R.id.button2);
22        startBtn.setOnClickListener(new mClick());
23        stopBtn.setOnClickListener(new mClick());
24    }
25
26    class mClick implements OnClickListener
27    {
28        @Override
29        public void onClick(View v)
30        {
31            if(v==startBtn)
32            {
33              startRecordAudio(path);
34            }
35            else if(v==stopBtn)
36            {
37              stopRecord();
38            }
39        }
40    }
41
42    void startRecordAudio(String path)
43    {
44        mRecorder=new MediaRecorder();
45        mRecorder.setAudioSource(MediaRecorder.AudioSource.MIC);    ← 设置音频源
46     mRecorder.setOutputFormat(
47                MediaRecorder.OutputFormat.THREE_GPP);    ← 设置输出格式
48        mRecorder.setAudioEncoder(
49                MediaRecorder.AudioEncoder.AMR_NB);    ← 设置编码格式
50        mRecorder.setOutputFile(path);    ← 设置输出文件路径
51        try {
52            mRecorder.prepare();    ← 准备录制
53            }catch (Exception e) {
54              System.out.println("Recorder err ... ");
55            }
56            mRecorder.start();    ← 开始录制
57    }
58
59    void stopRecord()    ← 结束录制
60    {
61        mRecorder.stop();    ← 停止录制
62        mRecorder.reset();    ← 重置
```

```
63            mRecorder.release();    ← 释放播放器的有关资源
64        }
65 }
```

（3）修改配置文件。在配置文件 AndroidManifest.xml 中增加音频捕获权限的语句。

- 音频捕获权限：

`<uses-permission android:name="android.permission.RECORD_AUDIO"/>`

- SD 卡的写操作权限：

`<uses-permission android:name="android.permission.WRITE_EXTERNAL_STORAGE"/>`

程序的运行结果如图 6.13 所示。

图 6.13 录音程序的运行界面

6.5.3 拍照

使用 android.hardware 包中的 Camera 类可以获取当前设备中的照相机服务接口，从而实现照相机的拍照功能。

1. 照片服务 Camera 类

Camera 类的常用方法如表 6-9 所示。

表 6-9 Camera 类的常用方法

方　　法	说　　明
open()	创建一个照相机对象
getParameters()	创建设置照相机参数的 Camera.Parameters 对象
setParameters(Camera.Parameters params)	设置照相机参数
setPreviewDisplay(SurfaceHolder holder)	设置取景预览
startPreview()	启动照片取景预览
stopPreview()	停止照片取景预览
release()	断开与照相机设备的连接，并释放资源
takePicture(Camera.ShutterCallback shutter, Camera.PictureCallback raw, Camera.PictureCallback jpeg)	进行照片拍摄

takePicture 方法有 3 个参数：
- 第 1 个参数 shutter 是关闭快门事件的回调接口；
- 第 2 个参数 raw 是获取照片事件的回调接口；
- 第 3 个参数 jpeg 也是获取照片事件的回调接口。

第 2 个参数与第 3 个参数的区别在于回调函数中传回的数据内容。第 2 个参数指定的回调函数中传回的数据内容是照片的原数据，而第 3 个参数指定的回调函数中传回的数据内容是已经按照 JPEG 格式进行编码的数据。

2. 实现拍照服务的主要步骤

在 Android 系统中，实现拍照服务的主要步骤如下：

（1）创建照相机对象。通过 Camera 类的 open()方法创建一个照相机对象：

```
Camera camera=Camera.open();
```

（2）设置参数。创建设置照相机参数的 Parameters 对象，并设置其相关参数：

```
parameters=mCamera.getParameters();
```

（3）对照片进行预览。通过照相机对象的 startPreview()方法和 stopPreview()方法启动或停止对照片的预览。

（4）拍摄照片。使用照相机接口的 takePicture()方法可以异步地进行照片拍摄。

通过照片事件的回调接口 PictureCallback，可以获取照相机所得到的图片数据，从而进行下一步的操作，例如将数据保存到本地存储、进行数据压缩、通过可视组件显示。

（5）停止拍摄

通过照相机对象的 release()方法可以断开与照相机设备的连接，并释放与该照相机接口有关的资源。

```
camera.release();
camera=null;
```

【例 6-10】设计一个简易照相机。

设计照相机，为了取景，需要应用 SurfaceView 组件来显示摄像头所能拍照的景物，然后使用回调接口 SurfaceHolder.Callback 监控取景视图。Callback 接口有 3 个方法需要实现：

视频演示

- surfaceCreated(SurfaceHolder holder)方法，用于初始化；
- surfaceChanged(SurfaceHolder holder, int format, int width,int height)方法，当景物发生变化时触发；
- surfaceDestroyed(SurfaceHolder holder)方法，释放对象时触发。

（1）创建用户界面。在界面设计中，设置两个按钮，分别为"拍照""退出"，然后设置一个 SurfaceView 组件用于取景预览，设置一个 ImageView 组件用于显示照片。其代码如下：

```
1    <LinearLayout xmlns:android="http://schemas.android.com/apk/res/android"
2        xmlns:tools="http://schemas.android.com/tools"
3        android:id="@+id/LinearLayout1"
4        android:layout_width="fill_parent"
```

```
5          android:layout_height="fill_parent"
6          android:orientation="vertical" >
7      <TextView
8          android:layout_width="wrap_content"
9          android:layout_height="wrap_content"
10         android:layout_gravity="center_horizontal"
11         android:padding="@dimen/padding_medium"
12         android:text="拍照测试"
13         android:textSize="20sp"
14         tools:context=".MainActivity" />
15     <LinearLayout
16         android:layout_width="fill_parent"
17         android:layout_height="wrap_content"
18         android:layout_gravity="center_horizontal"
19         android:gravity="center_horizontal" >
20         <Button
21             android:id="@+id/button1"
22             android:layout_width="110dp"
23             android:layout_height="wrap_content"
24             android:text="拍照" />
25         <Button
26             android:id="@+id/button2"
27             android:layout_width="110dp"
28             android:layout_height="wrap_content"
29             android:text="退出" />
30     </LinearLayout>
31     <ImageView                                      ← 用于显示拍摄的照片
32         android:id="@+id/imageView1"
33         android:layout_width="wrap_content"
34         android:layout_height="wrap_content" />
35     <SurfaceView                                    ← 用于取景预览
36         android:id="@+id/surfaceView1"
37         android:layout_width="320dp"
38         android:layout_height="240dp" />
39 </LinearLayout>
```

（2）设计控制文件 MainActivity.java 的代码如下：

```
1  package com.example.ex06_10;
2  import java.io.BufferedOutputStream;
3  import java.io.FileOutputStream;
4  import java.io.IOException;
5  import android.graphics.Bitmap;
6  import android.graphics.BitmapFactory;
7  import android.graphics.PixelFormat;
8  import android.hardware.Camera;
```

```
9    import android.hardware.Camera.PictureCallback;
10   import android.os.Bundle;
11   import android.util.Log;
12   import android.view.SurfaceHolder;
13   import android.view.SurfaceView;
14   import android.view.View;
15   import android.view.View.OnClickListener;
16   import android.widget.Button;
17   import android.widget.ImageView;
18   import android.app.Activity;
19
20   public class MainActivity extends Activity
21           implements SurfaceHolder.Callback     ◀── 应用 Callback 接口处理取景预览
22   {
23     Camera mCamera=null;
24     SurfaceView surfaceView;
25     SurfaceHolder holder;
26     ImageView mImageView;
27     Button cameraBtn, exitBtn;
28     String path="/sdcard/test/camera.jpg";      ◀── 定义存放照片的路径及文件名
29     @Override
30     public void onCreate(Bundle savedInstanceState)
31     {
32         super.onCreate(savedInstanceState);
33         setContentView(R.layout.activity_main);
34         mImageView=(ImageView)findViewById(R.id.imageView1);
35         cameraBtn=(Button)findViewById(R.id.button1);
36         exitBtn=(Button)findViewById(R.id.button2);
37         cameraBtn.setOnClickListener(new mClick());
38         exitBtn.setOnClickListener(new mClick());
39         surfaceView =(SurfaceView)findViewById(R.id.surfaceView1);
40         //创建 SurfaceHolder 对象
41         holder = surfaceView.getHolder();
42         //注册回调监听器
43         holder.addCallback(this);
44         //设置 SurfaceHolder 的类型
45         holder.setType(SurfaceHolder.SURFACE_TYPE_PUSH_BUFFERS);
46     }
47     class mClick implements OnClickListener
48     {
49       @Override
50       public void onClick(View v)
51       {
52         if(v == cameraBtn)
53             /*拍照并显示照片*/
```

```
54            mCamera.takePicture(null, null, new jpegCallback());   ← 拍照操作
55        else if(v==exitBtn)
56           exit();       ← 退出
57       }
58    }
59    void exit()
60    {
61       mCamera.release();
62       mCamera=null;
63    }
64    @Override
65    public void surfaceChanged(SurfaceHolder holder,   ← 当景物发生变化时触发
66                   int format, int width, int height)
67    {
68       /*调用设置照相机取景参数的方法*/
69       initCamera();
70    }
71    @Override
72    public void surfaceCreated(SurfaceHolder holder)
73    {
74       /*打开照相机*/
75       mCamera = Camera.open();
76       try {
77            /*设置预览*/
78            mCamera.setPreviewDisplay(holder);
79       } catch (IOException e) {
80            System.out.println("预览错误");
81       }
82    }
83    @Override
84    public void surfaceDestroyed(SurfaceHolder holder)
85    {        }
86   /*设置照相机取景参数*/
87    private void initCamera()
88    {
89       /*创建 Camera.Parameters 对象*/
90       Camera.Parameters parameters = mCamera.getParameters();
91       /*设置照片为 JPEG 格式*/
92       parameters.setPictureFormat(PixelFormat.JPEG);
93       /*指定 preview 的屏幕大小*/
94       parameters.setPreviewSize(320, 240);
95       /*设置图片分辨率大小*/
96       parameters.setPictureSize(320, 240);
```

```
97      /*将 Camera.Parameters 的设置作用于 Camera 对象*/
98      mCamera.setParameters(parameters);
99      /*打开预览*/
100     mCamera.startPreview();          ← 取景预览,此时还没有拍照保存为照片
101   }
102   /*通过 PictureCallback 接口进一步处理照相机所得到的图像数据*/
103   class jpegCallback implements PictureCallback
104   {
105    /**应用 onPictureTaken()方法将图像转换成 JPEG 格式后保存并预览照片,
106     * 其中,第一个参数 data 为存放照片数据的字节数组,
107     * 第二个参数 camera 为照片对象*/
108     @Override
109     public void onPictureTaken(byte[] data, Camera camera)
110     {
111       Bitmap bitmap=
112           BitmapFactory.decodeByteArray(data, 0, data.length);   ← 建立图像对象
113       try{
114          BufferedOutputStream outStream=new
115          BufferedOutputStream(new FileOutputStream(path));        ← 建立输出流对象
116          /*采用压缩转档方法*/
117          bitmap.compress(Bitmap.CompressFormat.JPEG, 80, outStream);
118          outStream.flush();          ← 调用 flush()方法更新 BufferStream
119          outStream.close();
120          /*显示拍摄的图像*/
121          mImageView.setImageBitmap(bitmap);
122       }
123       catch(Exception e)
124       {
125          Log.e("err", e.getMessage());
126       }
127     }
128   }
129 }
```

（3）修改配置文件 AndroidManifest.xml。在配置文件 AndroidManifest.xml 中增加允许操作 SD 卡和使用摄像头设备的语句：

```
<uses-permission android:name="android.permission.CAMERA"/>
<uses-permission
  android:name="android.permission.WRITE_EXTERNAL_STORAGE"/>
<uses-feature android:name="android.hardwre.camera"/>
<uses-feature android:name="android.hardwre.camera.autofocus"/>
```

程序的运行结果如图 6.14 所示。

图 6.14 简易照相机

6.6 将文本转换成语音

将文本转换成语音，使用 Android 的 TextToSpeech 组件可以方便地将其嵌入到游戏或者应用程序中，增强用户体验。该组件的功能很强大，并且利用它编写应用程序很简单。

文本转换语音 TextToSpeech 类（简称 TTS）的常用方法见表 6-10。

表 6-10 文本转换语音 TextToSpeech 类的常用方法

方　　法	功　　能
TextToSpeech(Context context, TextToSpeech.OnInitListener listener)	构造方法，创建一个文本转语音对象
speak(String text, int queueMode, HashMap<String, String> params)	实现文本转换语音功能，其参数 text 为转换文本，queueMode 为转换模式，其取值为 QUEUE_FLUSH 或 QUEUE_ADD，params 为 NULL
addSpeech(String text, String filename)	使文本 text 与声音文件 filename 建立映射
isSpeaking()	检查 TTS 引擎是否正在转换
shutdown()	停止转换，释放 TextToSpeech 引擎

在应用 TextToSpeech 设计文本转换语音程序时，还要实现 OnInitListener 接口，并覆盖 onInit(int status) 方法。在 onInit(int status) 方法中，对语音引擎进行初始化设置。

【例 6-11】应用 TextToSpeech 设计文本转换语音程序。

（1）创建用户界面。新建工程 ex06_11，修改 res\layout/activity_main.xml 文件，在其中添加一个文本编辑框（EditTex）和一个按钮（Button）。完整的 activity_main.xml 文件的代码如下：

视频演示

```
1  <?xml version="1.0" encoding="utf-8"?>
2  <LinearLayout xmlns:android="http://schemas.android.com/apk/res/android"
3      android:layout_width="fill_parent"
4      android:layout_height="fill_parent"
5      android:orientation="vertical" >
6      <TextView
7          android:layout_width="fill_parent"
```

```
8          android:layout_height="wrap_content"
9          android:text="@string/text"/>
10    <EditText
11         android:id="@+id/edit1"
12         android:layout_width="fill_parent"
13         android:layout_height="wrap_content" />
14    <Button
15         android:id="@+id/speak1"
16         android:layout_width="wrap_content"
17         android:layout_height="wrap_content"
18         android:text="@string/speak"   />
19 </LinearLayout>
```

（2）设计控制文件 MainActivity.java 的代码如下：

```
1  package com.ex06_11;
2  import android.app.Activity;
3  import android.os.Bundle;
4  import android.speech.tts.TextToSpeech;
5  import android.speech.tts.TextToSpeech.OnInitListener;
6  import android.util.Log;
7  import android.view.View;
8  import android.view.View.OnClickListener;
9  import android.widget.Button;
10 import android.widget.EditText;
11 public class MainActivity extends Activity
12 implements OnClickListener, OnInitListener
13 {
14     TextToSpeech tts;
15     int i=0;
16     @Override
17     public void onCreate(Bundle savedInstanceState)
18     {
19         super.onCreate(savedInstanceState);
20         setContentView(R.layout.main);
21         tts = new TextToSpeech(this, this);
22         Button speakButton = (Button)findViewById(R.id.speak1);
23         speakButton.setOnClickListener(this);    ← 监听按钮事件
24     }
25     @Override    ← 初始化
26     public void onInit(int status)
27     {
28         if (status==TextToSpeech.SUCCESS)
29         {
30             tts.speak("I'm ready!", TextToSpeech.QUEUE_FLUSH, null);
31         } else {
32             System.out.println("tts err!");
33         }
34 }
```

```
35  @Override
36  public void onClick(View v)      ◄── 处理 TextToSpeech 转换
37  {
38      EditText txt=(EditText)findViewById(R.id.edit1);
39      String str=txt.getText().toString();
40      tts.speak(str, TextToSpeech.QUEUE_FLUSH, null);  ◄── 调用语音引擎的方法
41  }
42  protected void onDestroy()       ◄── 关闭转换
43  {
44      super.onDestroy();
45      tts.shutdown();
46  }
47  }
```

程序运行结果如图 6.15 所示。其中，在文本框中输入要发音朗读的文字内容。

由于 Android 系统默认安装的 TTS 是 Pico TTS，它不支持中文，因此，需要下载安装另外的第三方语音包，通常可以使用 Svox、eSpeak 等。下载安装后打开文字转语音设置，选择 Svox Classic TTS 复选框，其语言选择中文。经设置后，TextToSpeech 组件就可以使用中文普通话发音了。

Svox 语音包图标如图 6.16 所示。

图 6.15　文本转语音　　　　　　　图 6.16　文本转语音的 Svox 语音包图标

6.7　图像处理技术

在程序设计过程中有时需要对图片做特殊的处理，例如将图片做出黑白或者老照片的效果，有时还要对图片进行变换，如拉伸、扭曲等。

这些效果在 Android 中有很好的支持，通过颜色矩阵（ColorMatrix）和坐标变换矩阵（Matrix）可以完美地做出上面的效果。

6.7.1　处理图像的颜色矩阵

1. 矩阵变换处理图像颜色原理

在 Android 系统中可以通过颜色矩阵来表示和处理图像的颜色，颜色矩阵是一个 4×5 的矩阵。

设颜色矩阵 A，第 1 行表示 R 红色分量，第 2 行表示 G 绿色分量，第 3 行表示 B 蓝色

分量，第 4 行表示透明度；颜色矩阵的第 5 列表示各个颜色的偏移量，如图 6.17 所示。

$$A = \begin{bmatrix} a & b & c & d & e \\ f & g & h & i & j \\ k & l & m & n & o \\ p & q & r & s & t \end{bmatrix} \qquad C = \begin{bmatrix} R \\ G \\ B \\ 1 \end{bmatrix} \qquad R = A * C = \begin{bmatrix} R' \\ G' \\ B' \\ A' \end{bmatrix}$$

（颜色矩阵） （颜色分量） （矩阵乘法运算）

图 6.17 颜色矩阵及运算

颜色矩阵在内存中是以一维数组的方式存储的，其格式如下：

[a, b, c, d, e, f, g, h, i, j, k, l, m, n, o, p, q, r, s, t]

在数字图像处理中，通过 R、G、B、A 4 个通道可以操作对应的颜色，从而做出各种特殊的颜色效果。颜色矩阵可以用来方便地修改 R、G、B、A 各分量的值，达到控制各颜色通道变化的目的。

颜色矩阵的运算规则是，矩阵 A 的一行乘以矩阵 C 的一列作为矩阵 R 的一行，C 矩阵是图像中包含的颜色分量 A、R、G、B 的信息，R 矩阵是用颜色矩阵应用于 C 之后的新的颜色分量，运算结果如下：

```
R' = a*R + b*G + c*B + d*A + e;    ← 红色分量
G' = f*R + g*G + h*B + i*A + j;    ← 绿色分量
B' = k*R + l*G + m*B + n*A + o;    ← 蓝色分量
A' = p*R + q*G + r*B + s*A + t;    ← 透明度
```

颜色矩阵并不复杂，需要使用的参数其实很少，而且很有规律：第 1 行决定红色，第 2 行决定绿色，第 3 行决定蓝色，第 4 行决定透明度，第 5 列是颜色的偏移量。

1）保持原有颜色不变的颜色矩阵

$$A = \begin{pmatrix} 1 & 0 & 0 & 0 & 0 \\ 0 & 1 & 0 & 0 & 0 \\ 0 & 0 & 1 & 0 & 0 \\ 0 & 0 & 0 & 1 & 0 \end{pmatrix}$$

如果把这个矩阵作用于各颜色分量，$R=A*C$，计算后会发现，各个颜色分量实际上没有任何改变（R' = R　G' = G　B' = B　A' = A）。

2）改变颜色分量

$$A = \begin{pmatrix} 1 & 0 & 0 & 0 & 100 \\ 0 & 1 & 0 & 0 & 100 \\ 0 & 0 & 1 & 0 & 0 \\ 0 & 0 & 0 & 1 & 0 \end{pmatrix}$$

这个矩阵红色分量增加 100，绿色分量增加 100，这样的效果就是图像偏黄，因为红色和绿色混合后得到黄色，黄色增加了 100，达到了泛黄的旧照片效果。

2. 颜色矩阵的常用方法

颜色矩阵的常用方法如表 6-11 所示。

表 6-11　颜色矩阵的常用方法

方　　法	说　　明
set(float[] src)	将 src 数组的值赋给颜色矩阵 A
float[] getArray()	返回颜色矩阵 A 的具体数值，以一阶数组形式表示
setScale (float rScale, float gScale, float bScale, float aScale)	设置矩阵的 R、G、B、A 变量的对应倍数，改变图像的亮度
setRotate (int axis, float degrees)	设置颜色分量旋转：axis==0 对应红色；axis==1 对应绿色；axis==2 对应蓝色
setSaturation (float sat)	通过改变矩阵的值设置图像的饱和度，参数 0 对应灰色图像，1 对应没有改变

【例 6-12】应用颜色矩阵变换改变图像颜色。

在本例中，从屏幕界面输入颜色矩阵数据，通过矩阵变换改变图像的颜色。其工作过程如下：

（1）从文本编辑框中获取输入的数据，将数据转换为浮点类型后，存放到数组 carray 中；

（2）用数组 carray 构建颜色矩阵；

（3）通过颜色矩阵对颜色通道过滤，然后用画布绘制一个新的图像。

在图形界面布局文件中，设置了一个自定义的视图 MyView，其 id 为 "imageView1"。然后设置了 20 个文本编辑框 editText1～editText20，用于输入颜色矩阵的数据值，为了直观，定义一个表格布局 TableLayout，将 20 个文本编辑框按 4 行 5 列的形式排列。XML 文件中设置的组件如图 6.18 所示。

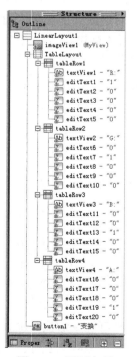

图 6.18　界面布局

视频演示

布局文件中的关键代码如下：

```xml
<LinearLayout xmlns:android="http://schemas.android.com/apk/res/android"
    …>
        <com.example.imagetest_02.MyView
            android:id="@+id/imageView1"
            android:layout_width="wrap_content"
            android:layout_height="246dp"
            android:layout_gravity="center_horizontal"
            android:layout_weight="1" />
    <TableLayout
        android:layout_width="match_parent"
        android:layout_height="wrap_content"
        android:shrinkColumns="1,2,3,4,5" >
        <TableRow
            android:id="@+id/tableRow1"
            android:layout_width="wrap_content"
            android:layout_height="wrap_content" >
            <TextView
                android:id="@+id/textView1"
                android:layout_width="wrap_content"
                android:layout_height="wrap_content"
                android:text="R:"
                android:textSize="20sp" />
            <EditText
                android:id="@+id/editText1"
                android:layout_width="wrap_content"
                android:layout_height="wrap_content"
                android:ems="5"
                android:text="1" >
            …         ← 在此省略了类似的<TableRow>和<EditText>语句
        </TableRow>
    </TableLayout>
    <Button
        android:id="@+id/button1"
        android:layout_width="wrap_content"
        android:layout_height="wrap_content"
        android:layout_gravity="center_horizontal|bottom"
        android:text="变换"
        android:textSize="20sp" />
</LinearLayout>
```

在 MyView 类中，定义了一个 setValues()方法，将 20 个文本编辑框中输入的数据赋给颜色矩阵，并重定义了 onDraw()方法，根据输入的数据设置颜色矩阵，再调用 Canvas 绘制新的图像。代码如下：

```java
1   package com.example.ex06_12;
2   import android.content.Context;
3   import android.graphics.Bitmap;
4   import android.graphics.BitmapFactory;
5   import android.graphics.Canvas;
6   import android.graphics.ColorMatrix;
7   import android.graphics.ColorMatrixColorFilter;
8   import android.graphics.Paint;
9   import android.util.AttributeSet;
10  import android.view.View;
11
12  public class MyView extends View
13  {
14      Paint mPaint=new Paint(Paint.ANTI_ALIAS_FLAG);
15      Bitmap mBitmap;
16      float [] array=new float[20];
17    public MyView(Context context, AttributeSet attrs)
18    {
19      super(context, attrs);
20      mBitmap=BitmapFactory.decodeResource(context.getResources(),
21                      R.drawable.like);
22    }
23    public void setValues(float []a)
24    {
25      for(int i=0;i<20;i++)
26         array[i]=a[i];
27    }
28    protected void onDraw(Canvas canvas)
29    {
30      super.onDraw(canvas);
31      Paint paint = mPaint;
32      paint.setColorFilter(null);
33      canvas.drawBitmap(mBitmap, 0, 0, paint);
34      ColorMatrix cMatrix = new ColorMatrix();    ← 实例化颜色矩阵对象
35      //设置颜色矩阵
36      cMatrix.set(array);    ← 将输入的数据构造颜色矩阵
37      //颜色滤镜,将颜色矩阵应用于图片
38    paint.setColorFilter(new ColorMatrixColorFilter(cMatrix));  ← 颜色矩阵过滤
39      //绘图
40      canvas.drawBitmap(mBitmap, 0, 0, paint);    ← 绘制经颜色矩阵变换后的新图像
41    }
42  }
```

主控文件比较简单,主要完成变量声明和初始化等任务。代码如下:

```
1   package com.example.ex06_12;
```

```java
2   import android.os.Bundle;
3   import android.view.View;
4   import android.view.View.OnClickListener;
5   import android.widget.Button;
6   import android.widget.EditText;
7   import android.app.Activity;
8
9   public class MainActivity extends Activity
10  {
11      EditText [] edit=new EditText[20];
12      int data[]={
13              R.id.editText1, R.id.editText2, R.id.editText3,
14              R.id.editText4, R.id.editText5, R.id.editText6,
15              R.id.editText7, R.id.editText8, R.id.editText9,
16              R.id.editText10, R.id.editText11, R.id.editText12,
17              R.id.editText13, R.id.editText14, R.id.editText15,
18              R.id.editText16, R.id.editText17, R.id.editText18,
19              R.id.editText19, R.id.editText20  };
20      float []carray=new float[20];
21      Button changeBtn;
22      MyView myView;
23      @Override
24      public void onCreate(Bundle savedInstanceState)
25      {
26          super.onCreate(savedInstanceState);
27          setContentView(R.layout.activity_main);
28          myView=(MyView)findViewById(R.id.imageView1);
29          changeBtn=(Button)findViewById(R.id.button1);
30          for(int i=0; i<20; i++)
31          {
32            edit[i]=(EditText)findViewById(data[i]);
33            carray[i]=Float.valueOf(edit[i].getText().toString());    ◀── 读取初始数据
34          }
35          changeBtn.setOnClickListener(new mClick());
36      }
37      class mClick implements OnClickListener
38      {
39        @Override
40        public void onClick(View v)
41        {
42            getValues();
43            myView.setValues(carray);    ◀── 将输入的数据构置颜色矩阵
44            myView.invalidate();
45        }
46      void getValues()
```

```
47      {
48          for(int i=0;i<20;i++)
49              carray[i]=
50                  Float.valueOf(edit[i].getText().toString());    ← 读取输入的数据
51      }
52  }
53 }
```

程序的运行结果如图 6.19 所示，改变颜色矩阵的值，经变换后，颜色发生改变。

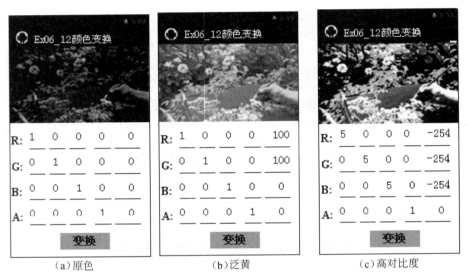

图 6.19　颜色矩阵改变颜色示例

6.7.2　处理图像的坐标变换矩阵

1. 矩阵变换处理图像原理

在 Android 系统中，可以通过坐标变换矩阵类来表示和处理图像。坐标变换矩阵是一个 3×3 的矩阵，图形的放大、缩小、移动、旋转、扭曲等效果都可以用坐标变换矩阵来完成。

设有坐标变换矩阵 A、原图像坐标矩阵 C，经坐标变换后，得到 R 矩阵，如图 6.20 所示。

$$A = \begin{bmatrix} a & b & c \\ d & e & f \\ 0 & 0 & i \end{bmatrix} \qquad C = \begin{bmatrix} x_0 \\ y_0 \\ 1 \end{bmatrix} \qquad R = A * C = \begin{bmatrix} x' \\ y' \\ 1 \end{bmatrix}$$

（坐标变换矩阵）　　　（坐标分量）　　　（矩阵乘法运算）

图 6.20　矩阵变换

坐标变换矩阵 A 的作用是对坐标（x_0, y_0）进行变换，其计算结果如下：

$$x' = a*x_0 + b*y_0 + c$$
$$y' = d*x_0 + e*y_0 + f$$

由于图片是由点阵和每一点上的颜色信息组成的，图形经坐标变换后，原来的坐标点被移动到新的坐标位置，从而发生图形的放大、缩小、移动、旋转、扭曲等效果。

1）平移变换

设图像原坐标为（x_0, y_0），经移动后到（x, y）位置，如图 6.21 所示。

其表达式为：

$$x = x_0 + \Delta x$$
$$y = y_0 + \Delta y$$

用矩阵表示为：

$$\begin{pmatrix} x \\ y \\ 1 \end{pmatrix} = \begin{pmatrix} 1 & 0 & \Delta x \\ 0 & 1 & \Delta y \\ 0 & 0 & 1 \end{pmatrix} * \begin{pmatrix} x_0 \\ y_0 \\ 1 \end{pmatrix}$$

（新坐标）　　（坐标变换矩阵）　　（原坐标）

平移变换效果如图 6.25（a）所示。

2）缩放变换

在坐标变换矩阵 $\begin{pmatrix} 1 & 0 & 0 \\ 0 & 1 & 0 \\ 0 & 0 & scale \end{pmatrix}$ 中，第 3 行第 3 列的元素 scale 是缩放的比例因子。

scale 通常取值为 1，表示图像大小不变，当 scale = 2 时表示图像缩小 1/2，scale = 0.5 时表示图像放大两倍。缩放变换效果如图 6.25（b）所示。

3）旋转变换

设图像原坐标为（x_0, y_0），经移动后到（x, y）位置，假定图像与坐标原点的距离为 r，如图 6.22 所示。

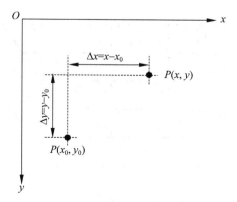

图 6.21　平移变换

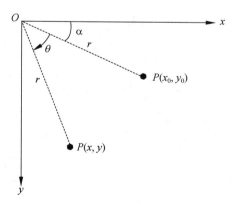

图 6.22　旋转变换

则

$$x_0 = x_0 \cdot \cos\theta - y_0 \cdot \sin\theta$$
$$y_0 = x_0 \cdot \sin\theta + y_0 \cdot \cos\theta$$

用矩阵表示为：

$$\begin{pmatrix} x \\ y \\ 1 \end{pmatrix} = \begin{pmatrix} \cos\theta & -\sin\theta & 0 \\ \sin\theta & \cos\theta & 0 \\ 0 & 0 & 1 \end{pmatrix} * \begin{pmatrix} x_0 \\ y_0 \\ 1 \end{pmatrix}$$

当 $\theta = 90°$ 时，

$$\begin{pmatrix} x \\ y \\ 1 \end{pmatrix} = \begin{pmatrix} 0 & -1 & 0 \\ 1 & 0 & 0 \\ 0 & 0 & 1 \end{pmatrix} * \begin{pmatrix} x_0 \\ y_0 \\ 1 \end{pmatrix}$$

即图像旋转 90°。旋转变换效果如图 6.25（c）所示。

4）对称变换

所谓对称变换，就是经过变化后的图像和原图像是关于某个轴对称的。例如，某点 P (x_0, y_0) 经过对称变换后得到 $P'(x, y)$，如图 6.23 所示。

用矩阵表示为：

$$\begin{pmatrix} x \\ y \\ 1 \end{pmatrix} = \begin{pmatrix} -1 & 0 & 0 \\ 0 & 1 & 0 \\ 0 & 0 & 1 \end{pmatrix} * \begin{pmatrix} x_0 \\ y_0 \\ 1 \end{pmatrix}$$

对称变换效果如图 6.25（d）、（e）所示。

5）错切变换

错切变换又称剪切变换，错切变换的效果就是让所有点的 x 坐标（或者 y 坐标）保持不变，而对应的 y 坐标（或者 x 坐标）按比例发生平移，且平移的大小和该点到 X 轴（或 Y 轴）的垂直距离成正比，如图 6.24 所示。

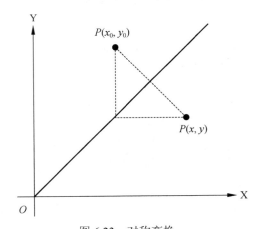

图 6.23　对称变换

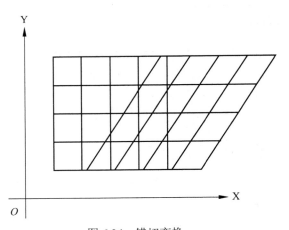

图 6.24　错切变换

用矩阵表示为：

$$\begin{pmatrix} x \\ y \\ 1 \end{pmatrix} = \begin{pmatrix} 1 & k & 0 \\ 0 & 1 & 0 \\ 0 & 0 & 1 \end{pmatrix} * \begin{pmatrix} x_0 \\ y_0 \\ 1 \end{pmatrix} \text{ 或 } \begin{pmatrix} x \\ y \\ 1 \end{pmatrix} = \begin{pmatrix} 1 & 0 & 0 \\ k & 1 & 0 \\ 0 & 0 & 1 \end{pmatrix} * \begin{pmatrix} x_0 \\ y_0 \\ 1 \end{pmatrix}$$

错切变换效果如图 6.25（f）所示。

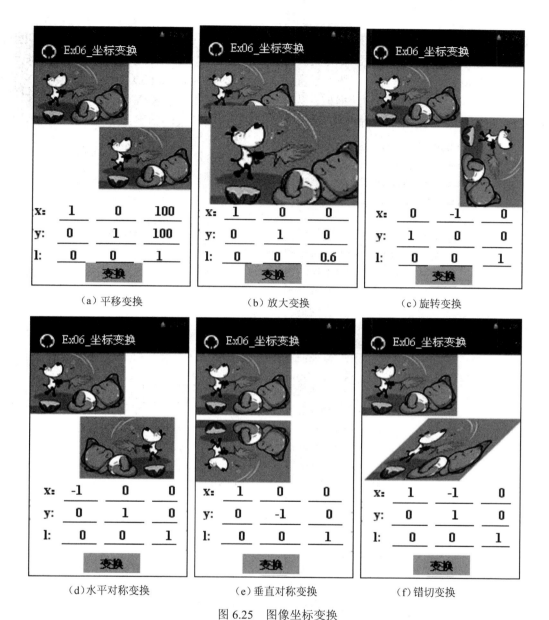

图 6.25 图像坐标变换

2. 坐标变换矩阵的常用方法

坐标变换矩阵类提供了许多变换方法，只要调用即可。其常用方法见表 6-12。

表 6-12 坐标变换矩阵的常用方法

方　　法	说　　明
setScale(float sx, float sy, float px, float py)	放大或缩小变换
setSkew(float kx, float ky, float px, float py)	斜切变换
setTranslate(float dx, float dy)	平移变换
setRotate(float degrees, float px, float py)	旋转变换

【例6-13】应用坐标矩阵变换改变图像大小及位置。

(1) 创建界面布局。在界面布局文件中,设置了一个自定义的视图 MyView,其 id 为 imageView1。然后设置了 9 个文本编辑框 editText1～editText9,用于输入变换矩阵的数据值,为了直观,定义了一个表格布局 TableLayout,将 9 个文本编辑框按 3 行 3 列的形式排列。其代码如下:

视频演示

```
1   <LinearLayout xmlns:android="http://schemas.android.com/apk/res/android"
2       android:id="@+id/LinearLayout1"
3       android:layout_width="fill_parent"
4       android:layout_height="fill_parent"
5       android:orientation="vertical" >
6       <com.example.ex06_13.MyView          ← 添加自定义视图 MyView
7           android:id="@+id/imageView1"
8           android:layout_width="wrap_content"
9           android:layout_height="246dp"
10          android:layout_weight="1" />
11      <TableLayout
12          android:layout_width="fill_parent"
13          android:layout_height="wrap_content"
14          android:shrinkColumns="1,2,3,4,5" >
15          <TableRow                         ← 表格布局第 1 行
16              android:id="@+id/tableRow1"
17              android:layout_width="wrap_content"
18              android:layout_height="wrap_content" >
19              <TextView
20                  android:id="@+id/textView1"
21                  android:layout_width="wrap_content"
22                  android:layout_height="wrap_content"
23                  android:text="X:"
24                  android:textSize="20sp" />
25              <EditText
26                  android:id="@+id/editText1"
27                  android:layout_width="wrap_content"
28                  android:layout_height="wrap_content"
29                  android:ems="3"
30                  android:text="1" >
31                  <requestFocus />
32              </EditText>
33              <EditText
34                  android:id="@+id/editText2"
35                  android:layout_width="wrap_content"
36                  android:layout_height="wrap_content"
37                  android:ems="3"
38                  android:text="0" />
```

```
39      <EditText
40          android:id="@+id/editText3"
41          android:layout_width="wrap_content"
42          android:layout_height="wrap_content"
43          android:ems="3"
44          android:text="0" />
45      </TableRow>
46      <TableRow              ←┤表格布局第2行│
47          android:id="@+id/tableRow2"
48          android:layout_width="wrap_content"
49          android:layout_height="wrap_content" >
50          <TextView
51              android:id="@+id/textView2"
52              android:layout_width="wrap_content"
53              android:layout_height="wrap_content"
54              android:text="Y:"
55              android:textSize="20sp" />
56          <EditText
57              android:id="@+id/editText4"
58              android:layout_width="wrap_content"
59              android:layout_height="wrap_content"
60              android:ems="5"
61              android:text="0" />
62          <EditText
63              android:id="@+id/editText5"
64              android:layout_width="wrap_content"
65              android:layout_height="wrap_content"
66              android:ems="6"
67              android:text="1" />
68          <EditText
69              android:id="@+id/editText6"
70              android:layout_width="wrap_content"
71              android:layout_height="wrap_content"
72              android:ems="5"
73              android:text="0" />
74      </TableRow>
75      <TableRow              ←┤表格布局第3行│
76          android:id="@+id/tableRow3"
77          android:layout_width="wrap_content"
78          android:layout_height="wrap_content" >
79          <TextView
80              android:id="@+id/textView3"
81              android:layout_width="wrap_content"
82              android:layout_height="wrap_content"
83              android:text="1:"
```

```
84              android:textSize="20sp" />
85          <EditText
86              android:id="@+id/editText7"
87              android:layout_width="wrap_content"
88              android:layout_height="wrap_content"
89              android:ems="5"
90              android:text="0" />
91          <EditText
92              android:id="@+id/editText8"
93              android:layout_width="wrap_content"
94              android:layout_height="wrap_content"
95              android:ems="5"
96              android:text="0" />
97          <EditText
98              android:id="@+id/editText9"
99              android:layout_width="wrap_content"
100             android:layout_height="wrap_content"
101             android:ems="5"
102             android:text="1" />
103     </TableRow>
104 </TableLayout>
105 <Button
106     android:id="@+id/button1"
107     android:layout_width="wrap_content"
108     android:layout_height="wrap_content"
109     android:layout_gravity="center_horizontal"
110     android:text="变换"
111     android:textSize="20sp" />
112 </LinearLayout>
```

（2）自定义视图 MyView.java 的代码如下：

```
1  package com.example.ex06_13;
2  import android.content.Context;
3  import android.graphics.Bitmap;
4  import android.graphics.BitmapFactory;
5  import android.graphics.Canvas;
6  import android.graphics.Matrix;
7  import android.graphics.Paint;
8  import android.util.AttributeSet;
9  import android.view.View;
10
11 public class MyView extends View
12 {
13   private Paint mPaint = new Paint(Paint.ANTI_ALIAS_FLAG);
14   private Bitmap mBitmap;
```

```java
15    private float [] array=new float[9];
16    public MyView(Context context, AttributeSet attrs)
17    {
18      super(context, attrs);
19      mBitmap = BitmapFactory.decodeResource(context.getResources(),
20                  R.drawable.boy);
21      invalidate();
22    }
23    public void setValues(float []a)
24    {
25      for(int i=0;i<9;i++)
26        array[i]=a[i];
27    }
28    protected void onDraw(Canvas canvas)
29    {
30       super.onDraw(canvas);
31      Paint paint = mPaint;
32      canvas.drawBitmap(mBitmap, 0, 0, paint);
33      Matrix matrix = new Matrix();
34      //为坐标变换矩阵设置响应的值
35      matrix.setValues(array);
36      //按照坐标变换矩阵的描述绘图
37      canvas.drawBitmap(mBitmap, matrix, paint);
38    }
39 }
```

（3）设计控制文件 MainActivity.java 的代码如下：

```java
1  package com.example.ex06_13;
2  import android.os.Bundle;
3  import android.app.Activity;
4  import android.view.View;
5  import android.view.View.OnClickListener;
6  import android.widget.Button;
7  import android.widget.EditText;
8
9  public class MainActivity extends Activity
10 {
11    MyView zsmView;
12    Button btn;
13    EditText [] edit=new EditText[9];
14    float []carray=new float[9];
15    int data[] = {
16          R.id.editText1, R.id.editText2, R.id.editText3,
17          R.id.editText4, R.id.editText5, R.id.editText6,
```

```
18                  R.id.editText7, R.id.editText8, R.id.editText9,
19            };
20      @Override
21    public void onCreate(Bundle savedInstanceState)
22    {
23          super.onCreate(savedInstanceState);
24          setContentView(R.layout.activity_main);
25          zsmView=(MyView)findViewById(R.id.imageView1);
26          btn=(Button)findViewById(R.id.button1);
27          btn.setOnClickListener(new mClick());
28      }
29    public void getValues()
30    {
31      for(int i=0;i<9;i++)
32      {
33          edit[i]=(EditText)findViewById(data[i]);
34          carray[i]=Float.valueOf(edit[i].getText().toString());
35      }
36    }
37    class mClick implements OnClickListener
38    {
39      @Override
40      public void onClick(View arg0)
41      {
42          getValues();
43        zsmView.setValues(carray);
44        zsmView.invalidate();
45      }
46    }
47 }
```

程序的运行结果如图 6.25 所示，当改变坐标矩阵中的数值时，经变换后，图像发生变化。

习 题 6

1. 设计一个可以移动的小球，当小球被拖到一个小矩形块中时退出程序。
2. 设计一个手绘图形的画板。
3. 建立一个手写字体识别的字体库。
4. 设计一个具有选歌功能的音频播放器。
5. 为例 6-7 的视频播放器添加停止播放的功能。
6. 设计一个图片编辑器，该编辑器具有图像预览、缩放、变形、调整色彩等功能。

第 7 章 后台服务与系统服务技术

7.1 后台服务 Service

Android 系统的 Service 是一种类似于 Activity 的组件，但 Service 没有用户操作界面，也不能自己启动，其主要作用是提供后台服务调用。Service 不像 Activity 那样，当用户关闭应用界面就停止运行，Service 会一直在后台运行，除非另有明确命令其停止。

通常使用 Service 为应用程序提供一些只需在后台运行的服务，或不需要界面的功能，例如，从 Internet 下载文件、控制 Video 播放器等。

Service 的生命周期中只有 3 个阶段：onCreate、onStartCommand、onDestroy。其生命周期的方法见表 7-1。

表 7-1 Service 服务的常用方法

方 法	说 明
onCreate()	创建后台服务
onStartCommand (Intent intent, int flags, int startId)	启动一个后台服务
onDestroy()	销毁后台服务，并删除所有调用
sendBroadcast(Intent intent)	继承父类 Context 的 sendBroadcast()方法，实现发送广播机制的消息
onBind(Intent intent)	与服务通信的信道进行绑定，服务程序必须实现该方法
onUnbind(Intent intent)	撤销与服务信道的绑定

通常，Service 要在一个 Activity 中启动，调用 Activity 的 startService(Intent)方法启动 Service。若要停止正在运行的 Service，则调用 Activity 的 stopService(Intent)方法关闭 Service。方法 startService()和 stopService()均继承于 Activity 及 Service 共同的父类 android.content.Context。

一个服务只能创建一次，销毁一次，但可以开始多次，即 onCreate()和 onDestroy()方法只能被调用一次，而 onStartCommand()方法可以被调用多次。后台服务的具体操作一般应该放在 onStartCommand()方法里面。如果 Service 已经启动，当再次启动 Service 时则不调用 onCreate()而直接调用 onStartCommand()。

设计一个后台服务的应用程序，大致有以下几个步骤。

（1）创建 Service 的子类：
- 编写 onCreate()方法，创建后台服务；

- 编写 onStartCommand()方法，启动后台服务；
- 编写 onDestroy()方法，终止后台服务，并删除所有调用。

（2）创建启动和控制 Service 的 Activity：
- 创建 Intent 对象，建立 Activity 与 Service 的关联；
- 调用 Activity 的 startService(Intent)方法启动 Service 后台服务；
- 调用 Activity 的 stopService(Intent)方法关闭 Service 后台服务。

（3）修改配置文件 AndroidManifest.xml，在配置文件 AndroidManifest.xml 的<application>标签中添加以下代码：

```
<service android:enabled="true" android:name=".AudioSrv" />
```

【例 7-1】一个简单的后台音乐服务程序示例。

本例通过一个按钮启动后台服务，在服务程序中播放音乐文件，演示服务程序的创建、启动，然后通过另一按钮演示服务程序的销毁过程。新建项目 ex07_01 后，将一个音频文件 mtest1.mp3 复制到应用程序的 res/raw 目录下。

视频演示

（1）界面布局文件 activity_main.xml 的代码如下：

```
1   <?xml version="1.0" encoding="utf-8"?>
2   <LinearLayout xmlns:android="http://schemas.android.com/apk/res/android"
3       android:layout_width="fill_parent"
4       android:layout_height="fill_parent"
5       android:orientation="vertical" >
6       <TextView
7           android:id="@+id/text1"
8           android:layout_width="fill_parent"
9           android:layout_height="wrap_content"
10          android:text="@string/hello"
11          android:textSize="24sp"/>
12      <Button
13          android:id="@+id/butn1"
14          android:layout_width="wrap_content"
15          android:layout_height="wrap_content"
16          android:text="启动后台音乐服务程序"
17          android:textSize="24sp" />
18       <Button
19          android:id="@+id/butn2"
20          android:layout_width="wrap_content"
21          android:layout_height="wrap_content"
22          android:text="关闭后台音乐服务程序"
23          android:textSize="24sp" />
24  </LinearLayout>
```

（2）新建后台服务程序 AudioSrv.java，其代码如下：

```
1   package com.ex07_01;
2   import android.app.Service;
```

```java
3    import android.content.Intent;
4    import android.media.MediaPlayer;
5    import android.os.IBinder;
6    import android.widget.Toast;
7
8    public class AudioSrv extends Service
9    {
10      MediaPlayer play;
11      @Override
12      public IBinder onBind(Intent intent)
13      {
14         return null;
15      }
16      public void onCreate()
17      {
18        super.onCreate();
19        play=MediaPlayer.create(this, R.raw.mtest1);    ← 创建调用资源音乐文件的对象
20        Toast.makeText(this, "创建后台服务...", Toast.LENGTH_LONG).show();
21      }
22      public int onStartCommand(Intent intent, int flags, int startId)
23      {
24          super.onStartCommand(intent, flags, startId);
25          play.start();    ← 开始播放音乐
26        Toast.makeText(this, "启动后台服务程序，播放音乐...",
27                       Toast.LENGTH_LONG).show();
28          return START_STICKY;
29      }
30      public void onDestroy()
31      {
32        play.release();
33        super.onDestroy();
34        Toast.makeText(this, "销毁后台服务！", Toast.LENGTH_LONG).show();
35      }
36   }
```

（3）设计启动后台服务的主控程序 MainActivity.java 的代码如下：

```java
1    package com.ex07_01;
2    import android.app.Activity;
3    import android.content.Context;
4    import android.content.Intent;
5    import android.os.Bundle;
6    import android.view.View;
7    import android.view.View.OnClickListener;
8    import android.widget.Button;
9    import android.widget.TextView;
10
```

```
11   public class MainActivity extends Activity
12   {
13      Button startbtn, stopbtn;
14      Context context;
15      Intent intent;
16      static TextView txt;
17    @Override
18    public void onCreate(Bundle savedInstanceState)
19    {
20       super.onCreate(savedInstanceState);
21       setContentView(R.layout.main);
22       startbtn=(Button)findViewById(R.id.butn1);
23       stopbtn=(Button)findViewById(R.id.butn2);
24       startbtn.setOnClickListener(new mClick());
25       stopbtn.setOnClickListener(new mClick());
26       txt=(TextView)findViewById(R.id.text1);
27       intent=new Intent(MainActivity.this, AudioSrv.class);   ← 创建 Intent 对象
28    }
29    class mClick implements OnClickListener   ← 定义一个类实现监听接口
30    {
31      public void onClick(View v)
32      {
33        if(v==startbtn)
34        {
35          MainActivity.this.startService(intent);   ← 启动 Intent 关联的 Service
36          txt.setText("start service ...");
37        }
38        else if(v==stopbtn)
39        {
40          MainActivity.this.stopService(intent);   ← 终止后台服务
41        }
42      }
43    }
44   }
```

（4）修改配置文件 AndroidManifest.xml。在配置文件 AndroidManifest.xml 的<application>标签中添加以下代码：

```
<service android:enabled="true" android:name=".AudioSrv" />
```

修改后，完整的 AndroidManifest.xml 文件如下：

```
1   <?xml version="1.0" encoding="utf-8"?>
2   <manifest xmlns:android="http://schemas.android.com/apk/res/android"
3       package="com.ex07_01"
4       android:versionCode="1"
5       android:versionName="1.0" >
```

```
6       <uses-sdk android:minSdkVersion="15" />
7       <application
8           android:icon="@drawable/ic_launcher"
9           android:label="@string/app_name" >
10          <activity
11              android:name=".MainActivity"
12              android:label="@string/app_name" >
13              <intent-filter>
14                  <action android:name="android.intent.action.MAIN" />
15                  <category android:name="android.intent.category.LAUNCHER" />
16              </intent-filter>
17          </activity>
18          <!-- 添加 AudioSrv 服务程序 -->
19          <service android:enabled="true" android:name=".AudioSrv" />
20      </application>
21  </manifest>
```

程序的运行结果如图 7.1 所示。

图 7.1　启动后台服务程序

7.2　信息广播机制 Broadcast

Broadcast 是 Android 系统应用程序之间传递信息的一种机制。当系统之间需要传递某些信息时，不是通过单击按钮之类的组件来触发事件，而是由系统自身通过系统调用来引发事件。这种系统调用是由 BroadcastReceiver 类实现的，把这种系统调用称为广播。BroadcastReceiver 也就是"广播接收者"的意思，顾名思义，它用来接收来自系统和应用中的广播信息。

在 Android 系统中有很多广播信息，例如当开机时系统会产生一条广播信息，接收到这条广播信息就能实现开机启动服务的功能；当网络状态改变时系统会产生一条广播信息，接收到这条广播信息就能及时地做出提示和保存数据等操作；当电池电量改变时系统会产生一条广播信息，接收到这条广播信息就能在电量低时告知用户及时保存进度，等等。

实现广播和接收机制有以下 5 个步骤：

（1）创建 Intent 对象，设置 Intent 对象的 action 属性。这个 action 属性是接收广播数据的标识，只有注册了相同 action 属性的广播接收器才能收到发送的广播数据。

```
Intent intent=new Intent();
intent.setAction("abc");    ← 设置 Intent 对象的 action 属性值为 "abc"
```

（2）编写需要广播的信息内容，将需要播发的信息封装到 Intent 中，通过 Activity 或 Service 继承其父类 Context 的 sendBroadcast()方法将 Intent 广播出去。

```
intent.putExtra("hello", "这是广播信息!");    ← 用键值对方式封装广播信息内容
sendBroadcast(intent);
```

（3）编写一个继承 BroadcastReceiver 的子类作为广播接收器，该对象是接收广播信息并对信息进行处理的组件。在子类中要重写接收广播信息的 onReceive()方法。

```
class TestReceiver extends  BroadcastReceiver
 {
    @Override
    public void onReceive(Context context, Intent intent)
     {
        /* 接收广播信息并对信息作出响应的代码 */
     }
 }
```

（4）在配置文件 AndroidManifest.xml 中注册广播接收类。

```
<service  android:name=".TestReceiver">    ← 注册广播接收类
  <intent-filter>
    <action android:name="abc" />    ← action 属性值相同才能接收到广播数据
  </intent-filter>
</service>
```

（5）销毁广播接收器。Android 系统在执行 onReceive()方法时会启动一个程序计时器，在一定时间内，广播接收器的实例会被销毁。因此，广播机制不适合传递大数据量的信息。

【例 7-2】一个简单的信息广播程序示例。

（1）主控文件 MainActivity.java 的代码如下：

```
1   package com.ex07_02;
2   import android.app.Activity;
3   import android.content.Intent;
4   import android.os.Bundle;
5   import android.view.View;
6   import android.view.View.OnClickListener;
7   import android.widget.Button;
8   import android.widget.TextView;
```

视频演示

```
9
10  public class MainActivity extends Activity
11  {
12      static TextView txt;
13      @Override
14      public void onCreate(Bundle savedInstanceState)
15      {
16         super.onCreate(savedInstanceState);
17         setContentView(R.layout.main);
18         txt=(TextView)findViewById(R.id.txt1);
19         Button btn=(Button)findViewById(R.id.button01);
20         btn.setOnClickListener(new mClick());
21      }
22
23      class mClick implements  OnClickListener
24      {
25        @Override
26        public void onClick(View v)
27        {
28           Intent intent=new Intent();
29           intent.setAction("abc");            ← 设置action属性值
30           Bundle bundle=new Bundle();
31           bundle.putString("hello", "这是广播信息!");   ← 设置广播的消息内容
32           intent.putExtras(bundle);
33           sendBroadcast(intent);              ← 发送广播消息
34        }
35      }
36  }
```

（2）广播接收器 TestReceiver.java 的代码如下：

```
1  package com.ex07_02;
2  import android.content.BroadcastReceiver;
3  import android.content.Context;
4  import android.content.Intent;
5
6  public class TestReceiver extends BroadcastReceiver    ← 定义广播接收器
7  {
8      @Override
9      public void onReceive(Context context, Intent intent)
10     {
11        String str=intent.getExtras().getString("hello");   ← 取出接收的数据
12        MainActivity.txt.setText(str);      ← 显示接收的数据
13     }
14  }
```

(3) 配置文件 AndroidManifest.xml 的代码如下：

```xml
1  <?xml version="1.0" encoding="utf-8"?>
2  <manifest xmlns:android="http://schemas.android.com/apk/res/android"
3      package="com.ex07_02"
4      android:versionCode="1"
5      android:versionName="1.0" >
6      <uses-sdk android:minSdkVersion="15" />
7      <application
8          android:icon="@drawable/ic_launcher"
9          android:label="@string/app_name" >
10         <activity
11             android:name=".MainActivity"
12             android:label="@string/app_name" >
13             <intent-filter>
14                 <action android:name="android.intent.action.MAIN" />
15                 <category android:name="android.intent.category.LAUNCHER" />
16             </intent-filter>
17         </activity>
18         <!-- 注册对应的广播接收类 -->
19         <receiver android:name=".TestReceiver">
20             <intent-filter>
21                 <!--注册广播的action, 与setAction()设置的值相同 -->
22                 <action android:name="abc" />
23             </intent-filter>
24         </receiver>
25     </application>
26 </manifest>
```

程序的运行结果如图 7.2 所示。

为了识别 Intent 对象的 action，有时在 IntentFilter 中设置 Intent 对象的 action，而注册广播接收器的工作由 registerReceiver()方法完成。

图 7.2 简单的广播示例

registerReceiver(mBroadcast, filter)方法有两个参数，其中，参数 mBroadcast 是广播接收器 BroadcastReceiver 对象，filter 是 IntentFilter 对象。

【例 7-3】由一个后台服务广播音乐的播放或暂停信息，接收器接收到信息后，执行改变用户界面按钮上文本的操作。

在本例中，创建了 3 个类：MainActivity、AudioService 和 Broadcast，MainActivity 负责用户的交互界面，并启动后台服务；AudioService 是 Service 的子类，在后台提供播放音乐或暂停、停止音乐等工作，同时发送改变交互界面的广播信息；Broadcast 是 BroadcastReceiver 的子类，负责接收广播信息，更改交互界面。这 3 个类的工作流程如图 7.3 所示。

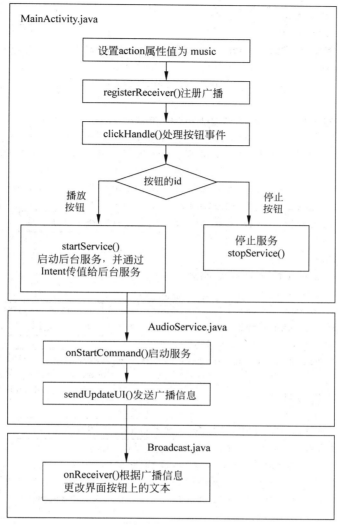

图 7.3 工作流程

（1）主控文件 MainActivity.java 的代码如下：

```
1   package com.ex07_03;
2   import android.app.Activity;
3   import android.content.Intent;
4   import android.content.IntentFilter;
5   import android.os.Bundle;
6   import android.view.View;
7   import android.widget.Button;
8
9   public class MainActivity extends Activity
10  {
11    Broadcast mBroadcast=null;
12    static Button btnStart;
```

视频演示

```
13    Button btnStop;
14    Intent intent;                              ← 事先将音乐文件
15    String AUDIO_PATH="/sdcard/music/mtest2.mp3";  复制到 SD 卡上
16     @Override
17    public void onCreate(Bundle savedInstanceState)
18    {
19        super.onCreate(savedInstanceState);
20        setContentView(R.layout.main);
21        btnStart=(Button) findViewById(R.id.btnPlayOrPause);
22        btnStop=(Button) findViewById(R.id.btnStop);     ← music 要在配置文
23        IntentFilter filter=new IntentFilter("music");     件 manifest.xml
24         mBroadcast=new Broadcast();                      中注册
25        registerReceiver(mBroadcast, filter);  ← 注册广播
26    }
27    @Override
28    protected void onDestroy()
29    {
30      super.onDestroy();                          ← 在 onDestroy()方法中解除注
31      unregisterReceiver(mBroadcast);               册，否则退出时会报异常
32    }
33    public void clickHandle(View v)  ← 处理按钮事件，在用户界面程序的按钮
34    {                                  的 onClick 属性中设置调用，按钮的属
35      switch(v.getId())                性为 android:onclick="clickHandle"
36      {
37      case R.id.btnPlayOrPause:  ← 资源中播放按钮的 id
38        intent = new Intent(MainActivity.this,
39             com.ex07_02.AudioService.class);      ← 绑定键值对，把音
40        Bundle bundle=new Bundle();                  乐文件的路径传
41        bundle.putString("audioPath", AUDIO_PATH); ← 递给后台服务
42        intent.putExtras(bundle);
43        startService(intent);  ← 启动服务
44        break;
45      case R.id.btnStop:  ← 资源中停止按钮的 id
46        if(intent != null)
47        {
48           stopService(intent);  ← 停止服务
49        }
50        break;
51      }
52    }
53  }
```

（2）后台服务程序 AudioService.java 的代码如下：

```
1   package com.ex07_03;
2   import android.app.Service;
3   import android.content.Intent;
```

```java
4    import android.media.MediaPlayer;
5    import android.os.Bundle;
6    import android.os.IBinder;
7    import android.util.Log;
8
9    public class  AudioService extends Service
10   {
11     private MediaPlayer mediaPlayer=null;
12     private Intent intent2=null;
13     private Bundle bundle2=null;
14     private String audioPath ;
15     @Override
16     public IBinder onBind(Intent intent)
17     {
18       return null;
19     }
20     public int onStartCommand(Intent intent, int flags, int startId)
21     {
22       super.onStartCommand(intent, flags, startId);
23       audioPath=intent.getExtras().getString("audioPath");
24       /** 1.正在播放
25        *    使其暂停播放,并通知界面将Button的值改为"播放"
26        *    (如果正在播放,Button值是"暂停")
27        */
28       if(mediaPlayer != null && mediaPlayer.isPlaying())
29       {
30          mediaPlayer.pause();
31          sendUpdateUI(1);   //更新界面
32       }
33       /**
34        *2.正在暂停
35        */
36       else{
37          if(mediaPlayer==null)
38          {
39             mediaPlayer=new MediaPlayer();   //如果被停止了,则为null
40             try {
41                mediaPlayer.reset();
42                mediaPlayer.setDataSource(audioPath) ;
43                mediaPlayer.prepare();
44                } catch (Exception e)
45                { Log.e("player", "player prepare() err");  }
46          }
47          mediaPlayer.start();
48          sendUpdateUI(0);    //更新界面
```

```
49            }
50         return START_STICKY;
51   }
52    @Override
53    public void onDestroy()
54    {
55        if(mediaPlayer !=null)
56        {
57            mediaPlayer.release();    //停止时要 release
58            sendUpdateUI(2);           //更新界面
59        }
60        super.onDestroy();
61    }
62    private void sendUpdateUI(int flag)
63    {
64        intent2=new Intent();
65        intent2.setAction("music");    ← action 属性值
66        bundle2=new Bundle();
67        bundle2.putInt("backFlag", flag);    ← 把 flag 传回去
68        intent2.putExtras(bundle2);
69        sendBroadcast(intent2);
70        /**发送广播
71         * 后台服务把键名为 backFlag 的信息广播出去
72         * 发送后,在 Activity 中的 updateUIReceiver 的 onReceiver()方法
73         * 就能做相应的更新界面的工作了
74         */
75    }
76  }
```

(3) 广播接收器 Broadcast.java 的代码如下：

```
1    package com.ex07_03;
2    import android.content.BroadcastReceiver;
3    import android.content.Context;
4    import android.content.Intent;
5    /*广播接收器*/
6    class Broadcast extends  BroadcastReceiver
7    {
8     @Override
9     public void onReceive(Context context, Intent intent)
10    {
11      /**
12       * 更新界面。这里改变 Button 的值
13       * 从 Intent 获取接收到的广播数据,
14       * 数据值为 0 表示此时是播放、为 1 表示是暂停、为 2 表示是停止
15       */
```

```
16      int backFlag=intent.getExtras().getInt("backFlag");
17      switch(backFlag)
18      {
19        case 0:
20          MainActivity.btnStart.setText("暂停");
21          break;
22        case 1:
23        case 2:
24          MainActivity.btnStart.setText("播放");
25          break;
26      }
27    }
28  }
```

（4）在配置文件 AndroidManifest.xml 中注册服务和广播的代码如下：

```
1   <?xml version="1.0" encoding="utf-8"?>
2   <manifest xmlns:android="http://schemas.android.com/apk/res/android"
3       package="com.ex07_03"
4       android:versionCode="1"
5       android:versionName="1.0" >
6       <uses-sdk android:minSdkVersion="15" />
7       <application
8           android:icon="@drawable/ic_launcher"
9           android:label="@string/app_name" >
10          <activity
11              android:name=".MainActivity"
12              android:label="@string/app_name" >
13              <intent-filter>
14                  <action android:name="android.intent.action.MAIN" />
15                  <category android:name="android.intent.category.LAUNCHER" />
16              </intent-filter>
17          </activity>
18          <service android:enabled="true" android:name=".AudioService" />
19              <intent-filter>
20                  <action android:name="music" />
21              </intent-filter>
22          </service>
23      </application>
24  </manifest>
```

程序的运行结果如图 7.4 所示。

图 7.4 运行后台广播服务

7.3 系统服务

在第 4 章学习了应用 Intent 进行参数传递,Android 系统提供了很多标准的系统服务,这些系统服务都可以简单地通过 Intent 进行广播。

7.3.1 Android 的系统服务

Android 系统提供了大量的标准系统服务,这些系统服务用于完成不同的功能,Android 的系统服务如表 7-2 所示。

表 7-2 Android 的系统服务

系 统 服 务	作 用
WINDOW_SERVICE ("window")	窗体管理服务
LAYOUT_INFLATER_SERVICE ("layout_inflater")	布局管理服务
ACTIVITY_SERVICE ("activity")	Activity 管理服务
POWER_SERVICE ("power")	电源管理服务
ALARM_SERVICE ("alarm")	时钟管理服务
NOTIFICATION_SERVICE ("notification")	通知管理服务
KEYGUARD_SERVICE ("keyguard")	键盘锁服务
LOCATION_SERVICE ("location")	基于地图的位置服务
SEARCH_SERVICE ("search")	搜索服务
VIBRATOR_SERVICE ("vibrator")	振动管理服务
CONNECTIVITY_SERVICE ("connection")	网络连接服务
WIFI_SERVICE ("wifi")	WiFi 连接服务
INPUT_METHOD_SERVICE ("input_method")	输入法管理服务
TELEPHONY_SERVICE ("telephony")	电话服务
DOWNLOAD_SERVICE ("download")	HTTP 的下载服务

系统服务实际上可以看作是一个对象,通过 Activity 类的 getSystemService 方法可以获得指定的对象(系统服务)。下面详细讲解几个常见的系统服务。

7.3.2 系统通知服务 Notification

Notification 是 Android 系统的一种通知服务,当手机来电、来短信、响闹钟铃声时,在状态栏中会显示相应通知的图标和文字,提示用户处理。当拖动状态栏时,可以查看这些信息。

Notification 提供了声音、振动等属性,表 7-3 列出了 Notification 的部分常见属性。

表 7-3 Notification 的部分常见属性

属　　性	说　　明
audioStreamType	所用音频流的类型
contentIntent	设置单击通知条目所执行的 Intent
contentView	设置状态栏显示通知的视图
defaults	设置成默认值
deleteIntent	删除通知所执行的 Intent
icon	设置状态栏上显示的图标
iconLevel	设置状态栏上显示图标的级别
ledARGB	设置 LED 灯的颜色
ledOffMS	设置关闭 LED 时的闪烁时间（以毫秒计算）
ledOnMS	设置开启 LED 时的闪烁时间（以毫秒计算）
sound	设置通知的声音文件
tickerText	设置状态栏上显示的通知内容
vibrate	设置振动模式
when	设置通知发生的时间

系统通知服务 Notification 由系统通知管理对象 NotificationManager 进行管理及发布通知，由 getSystemService(NOTIFICATION_SERVICE)创建 NotificationManager 对象：

```
NotificationManager n_Manager=
    (NotificationManager)getSystemService(NOTIFICATION_SERVICE);
```

NotificationManager 对象通过 notify(int id, Notification notification) 方法把通知发送到状态栏，通过 cancelAll()方法取消之前显示的所有通知。

【例 7-4】在状态栏显示系统通知服务的应用示例。

在界面布局文件中设置两个按钮，分别为"发送系统通知"和"删除通知"。设计控制文件 MainActivity.java 如下：

```
1   package com.example.ex07_04;
2   import android.os.Bundle;
3   import android.app.Activity;
4   import android.app.Notification;
5   import android.app.NotificationManager;
6   import android.app.PendingIntent;
7   import android.content.Intent;
8   import android.view.View;
9   import android.view.View.OnClickListener;
10  import android.widget.Button;
11
```

视频演示

```
12  public class MainActivity extends Activity
13  {
14    NotificationManager n_Manager;
15    Notification notification;
16    Button btn1, btn2;
17    @Override
18    public void onCreate(Bundle savedInstanceState)
19    {
20        super.onCreate(savedInstanceState);
21        setContentView(R.layout.activity_main);
22        String service=NOTIFICATION_SERVICE;
23        n_Manager=(NotificationManager)getSystemService(service);
24        btn1=(Button)findViewById(R.id.btn1);
25        btn1.setOnClickListener(new mClick());
26        btn2=(Button)findViewById(R.id.btn2);
27        btn2.setOnClickListener(new mClick());
28    }
29    class mClick implements OnClickListener
30    {
31        @Override
32      public void onClick(View arg0)
33      {
34        if(arg0==btn1)
35        {
36          int icon=R.drawable.ic_con;
37          CharSequence tickerText="紧急通知,程序已启动";
38          long when=System.currentTimeMillis();
39          notification=new Notification(icon, tickerText, when);
40          Intent intent=new Intent(MainActivity.this, MainActivity.class);
41          PendingIntent pi=PendingIntent.getActivity(MainActivity.this,
42                              0 , intent , 0);
43          notification.setLatestEventInfo(MainActivity.this,
44                              "通知", "通知内容", pi);
45          n_Manager.notify(0, notification);
46        }
47        else if(arg0==btn2)
48        {
49          n_Manager.cancelAll();
50        }
51      }
52    }
53  }
```

程序的运行结果如图 7.5 所示。

图 7.5 在状态栏显示系统通知

7.3.3 系统定时服务 AlarmManager

系统定时服务 AlarmManager 又称为系统闹钟服务。其作用是在到达设定的时间后，AlarmManager 广播一个 Intent 信息。AlarmManager 常用的属性和方法见表 7-4。

表 7-4　AlarmManager 常用的属性和方法

属性或方法名称	说　　明
ELAPSED_REALTIME	设置闹钟时间，从系统启动开始
ELAPSED_REALTIME_WAKEUP	设置闹钟时间，从系统启动开始，如果设备休眠则唤醒
INTERVAL_DAY	设置闹钟时间，间隔 24 小时
INTERVAL_FIFTEEN_MINUTES	间隔 15 分钟
INTERVAL_HALF_DAY	间隔 12 小时
INTERVAL_HALF_HOUR	间隔 30 分钟
INTERVAL_HOUR	间隔 1 小时
RTC	设置闹钟时间，从系统当前时间开始（System.currentTimeMillis()）
RTC_WAKEUP	设置闹钟时间，从系统当前时间开始，设备休眠则唤醒
set(int type,long tiggerAtTime, PendingIntent operation)	设置在某个时间执行闹钟
setRepeating(int type,long triggerAtTiem, long interval, PendingIntent operation)	设置在某个时间重复执行闹钟
setInexactRepeating(int type,long triggerAtTiem, long interval,PendingIntent operation)	设置在某个时间重复执行闹钟，而不是间隔固定时间
cancel(PendingIntent)	取消闹钟

AlarmManager 服务主要有两种应用：
- 在指定时长后执行某项操作；
- 周期性地执行某项操作。

下面通过示例说明这两种应用。

【例 7-5】AlarmManager 时钟服务示例。
其控制文件的代码如下：

视频演示

```java
1   package com.example.ex07_05;
2   import java.util.Calendar;
3   import android.os.Bundle;
4   import android.os.SystemClock;
5   import android.view.View;
6   import android.view.View.OnClickListener;
7   import android.widget.Button;
8   import android.widget.Toast;
9   import android.app.Activity;
10  import android.app.AlarmManager;
11  import android.app.PendingIntent;
12  import android.content.Intent;
13
14  public class MainActivity extends Activity
15  {
16    Button btn1, btn2, btn3;
17    Intent intent;
18    PendingIntent sender;
19    @Override
20    public void onCreate(Bundle savedInstanceState)
21    {
22        super.onCreate(savedInstanceState);
23        setContentView(R.layout.activity_main);
24        btn1=(Button)findViewById(R.id.button1);
25        btn1.setOnClickListener(new mClick());
26        btn2=(Button)findViewById(R.id.button2);
27        btn2.setOnClickListener(new mClick());
28        btn3=(Button)findViewById(R.id.button3);
29        btn3.setOnClickListener(new mClick());
30    }
31    class mClick implements OnClickListener
32    {
33    @Override
34    public void onClick(View v)
35    {
36        switch (v.getId())
37        {
38        case R.id.button1:
39            timing(); break;
40        case R.id.button2:
41            cycle(); break;
42        case R.id.button3:
43            cancel(); break;
44        }
45    }
46  }
47  /**
```

```
48      * 定时: 5秒后发送一个广播,广播接收后 Toast 提示定时操作完成
49      */
50     void  timing()
51     {
52        intent=new Intent(MainActivity.this, alarmreceiver.class);
53        intent.setAction("aaa");
54        sender=PendingIntent.getBroadcast(MainActivity.this, 0, intent, 0);
55        Calendar calendar=Calendar.getInstance();
56        calendar.setTimeInMillis(System.currentTimeMillis());
57        calendar.add(Calendar.SECOND, 5);      ← 设定一个5秒后的时间
58        AlarmManager alarm=(AlarmManager)getSystemService(ALARM_SERVICE);
59        alarm.set(AlarmManager.RTC_WAKEUP, calendar.getTimeInMillis(),
60                         sender);
61        Toast.makeText(MainActivity.this, "5秒后alarm开启",
62                         Toast.LENGTH_LONG).show();
63     }
64     /**
65      * 定义循环: 每5秒发送一个广播,广播接收后 Toast 提示定时操作完成
66      */
67     void cycle()
68     {
69        Intent intent=new Intent(MainActivity.this, alarmreceiver.class);
70        intent.setAction("repeating");
71        PendingIntent sender=PendingIntent.getBroadcast(MainActivity.this,
72                                          0, intent, 0);
73        /*开始时间*/
74        long firstime=SystemClock.elapsedRealtime();
75        AlarmManager am=(AlarmManager)getSystemService(ALARM_SERVICE);
76        /*5秒一个周期,不停地发送广播*/
77        am.setRepeating(AlarmManager.ELAPSED_REALTIME_WAKEUP ,
78                         firstime, 5*1000, sender);
79     }
80     /**
81      * 取消周期性地发送信息
82      */
83     void cancel()
84     {
85       Intent intent=new Intent(MainActivity.this, alarmreceiver.class);
86       intent.setAction("repeating");
87       PendingIntent sender=PendingIntent
88                  .getBroadcast(MainActivity.this, 0, intent, 0);
89       AlarmManager alarm=(AlarmManager)getSystemService(ALARM_SERVICE);
90       alarm.cancel(sender);
91     }
92  }
```

广播信息 alarmreceiver.java 代码如下:

```
1  package com.example.ex07_05;
2  import android.content.BroadcastReceiver;
3  import android.content.Context;
4  import android.content.Intent;
5  import android.widget.Toast;
6  public class alarmreceiver extends BroadcastReceiver
7  {
8    public void onReceive(Context context, Intent intent)
9    {
10   if(intent.getAction().equals("aaa"))
11      Toast.makeText(context, "时间到, 上课了! ", Toast.LENGTH_LONG).show();
12   else if(intent.getAction().equals("repeating"))
13      Toast.makeText(context, "时间到,起床了!",Toast.LENGTH_LONG).show();
14   }
15 }
```

在配置文件 AndroidManifest.xml 倒数第二行("</application>"的上一行)加入下列注册广播信息的语句:

```
1    <receiver  android:name="com.example.ex07_05.alarmreceiver">  <!--广播接收类 -->
2    <intent-filter>
3        <action android:name="aaa" />  <!--接受 广播注册的广播动作 -->
4    </intent-filter>
5    <intent-filter>
6        <action android:name="repeating" />
7    </intent-filter>
8    </receiver>
```

程序的运行结果如图 7.6 所示。

图 7.6 时钟服务示例

7.3.4 系统功能的调用

Android 系统通过 Intent 的 action 属性可以调用系统功能。常用的系统功能及调用语句见表 7-5。

表 7-5 常用的系统功能及调用语句

系统功能	调用语句

浏览网页	`Uri uri=Uri.parse("http://www.google.com");`
	`Intent it=new Intent(Intent.ACTION_VIEW,uri);`
	`startActivity(it);`

从 Google 搜索内容	`Intent intent=new Intent();`
	`intent.setAction(Intent.ACTION_WEB_SEARCH);`
	`intent.putExtra(SearchManager.QUERY,"searchString")`
	`startActivity(intent);`

显示地图	`Uri uri=Uri.parse("geo:38.899533,-77.036476");`
	`Intent it=new Intent(Intent.Action_VIEW,uri);`
	`startActivity(it);`

路径规划	`Uri uri=Uri.parse("http://maps.google.com/maps?f=dsaddr=startLat`
	`%20startLng&daddr=endLat%20endLng&hl=en");`
	`Intent it=new Intent(Intent.ACTION_VIEW,URI);`
	`startActivity(it);`

拨打电话	`Uri uri=Uri.parse("tel:xxxxxx");`
	`Intent it=new Intent(Intent.ACTION_DIAL,uri);`
	`startActivity(it);`

发送短信程序	`Intent it=new Intent(Intent.ACTION_VIEW);`
	`it.putExtra("sms_body", "TheSMS text");`
	`it.setType("vnd.android-dir/mms-sms");`
	`startActivity(it);`
	
	`Uri uri =Uri.parse("smsto:0800000123");`
	`Intent it=new Intent(Intent.ACTION_SENDTO, uri);`
	`it.putExtra("sms_body", "TheSMS text");`
	`startActivity(it);`
	
	`String body="this is sms demo";`
	`Intent mmsintent=new Intent(Intent.ACTION_SENDTO,`
	`                Uri.fromParts("smsto", number, null));`
	`mmsintent.putExtra(Messaging.KEY_ACTION_SENDTO_MESSAGE_BODY,body);`
	`mmsintent.putExtra(Messaging.KEY_ACTION_SENDTO_COMPOSE_MODE,true);`
	`mmsintent.putExtra(Messaging.KEY_ACTION_SENDTO_EXIT_ON_SENT,true);`
	`startActivity(mmsintent);`

发送 E-mail	`Uri uri=Uri.parse("mailto:xxx@abc.com");`
	`Intent it=new Intent(Intent.ACTION_SENDTO, uri);`
	`startActivity(it);`
	`Intent it=new Intent(Intent.ACTION_SEND);`
	`it.putExtra(Intent.EXTRA_EMAIL,"me@abc.com");`
	`it.putExtra(Intent.EXTRA_TEXT, "Theemail body text");`
	`it.setType("text/plain");`
	`startActivity(Intent.createChooser(it,"Choose Email Client"));`
	
	`Intent it=new Intent(Intent.ACTION_SEND);`

系统功能	调用语句

发送 E-mail	`String[] tos={"me@abc.com"};`
	`String[]ccs={"you@abc.com"};`
	`it.putExtra(Intent.EXTRA_EMAIL, tos);`
	`it.putExtra(Intent.EXTRA_CC, ccs);`
	`it.putExtra(Intent.EXTRA_TEXT, "Theemail body text");`
	`it.putExtra(Intent.EXTRA_SUBJECT, "Theemail subject text");`
	`it.setType("message/rfc822");`
	`startActivity(Intent.createChooser(it,"Choose Email Client"));`

发送邮件的附件	`Intent it=new Intent(Intent.ACTION_SEND);`
	`it.putExtra(Intent.EXTRA_SUBJECT, "Theemail subject text");`
	`it.putExtra(Intent.EXTRA_STREAM,"file:///sdcard/mysong.mp3");`
	`sendIntent.setType("audio/mp3");`
	`startActivity(Intent.createChooser(it,"Choose Email Client"));`

播放多媒体	`Intent it=new Intent(Intent.ACTION_VIEW);`
	`Uri uri=Uri.parse("file:///sdcard/song.mp3");`
	`it.setDataAndType(uri,"audio/mp3");`
	`startActivity(it);`
	`Uri uri=Uri.withAppendedPath(MediaStore.Audio.Media.`
	`        INTERNAL_CONTENT_URI,"1");`
	`Intent it=new Intent(Intent.ACTION_VIEW,uri);`
	`startActivity(it);`

打开录音机	`Intent mi=new Intent(Media.RECORD_SOUND_ACTION);` `startActivity(mi);`

【例 7-6】调用系统的短信发送功能示例。

本示例仅设置一个按钮，在按钮事件中添加发送短信的代码。

（1）编写调用系统短信发送功能的源程序，其代码如下：

视频演示

```
1   package com.example.ex07_06;
2   import android.net.Uri;
3   import android.os.Bundle;
4   import android.app.Activity;
5   import android.content.Intent;
6   import android.view.View;
7   import android.view.View.OnClickListener;
8   import android.widget.Button;
9
10  public class MainActivity extends Activity
11  {
12      Button btn_sms;
13      @Override
14      public void onCreate(Bundle savedInstanceState)
15      {
```

```
16        super.onCreate(savedInstanceState);
17        setContentView(R.layout.activity_main);
18        btn_sms=(Button)findViewById(R.id.btn1);
19        btn_sms.setOnClickListener(new mClick());
20    }
21    class mClick implements OnClickListener
22    {
23     @Override
24     public void onClick(View arg0)
25     {
26        Uri uri =Uri.parse("smsto:13900100100");
27        Intent it = new Intent(Intent.ACTION_SENDTO,uri);
28        it.putExtra("sms_body", "TheSMS text");
29        startActivity(it);
30     }
31    }
32 }
```

（2）在配置文件 AndroidManifest.xml 中添加访问网络权限的语句：

```
<uses-permission
        android:name="android.permission.INTERNET">
</uses-permission>
```

程序的运行结果如图 7.7 所示。

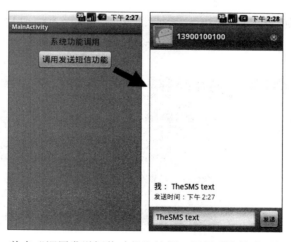

图 7.7 单击"调用发送短信功能"按钮，调用系统的发送短信功能

习 题 7

1. 结合例 7-1 和例 7-3，编写一个功能较完善的后台音乐播放器。
2. 编写一个短信服务平台。

第 8 章　　数 据 存 储

本章介绍 Android 系统的数据存储方法。Android 系统提供了多种数据存储方式，有 SQLite 数据库存储方式、文件存储方式、XML 文件的 SharedPreference 存储方式等。

8.1　SQLite 数据库

8.1.1　SQLite 数据库简介

SQLite 数据库是一个关系型数据库，因为它很小，引擎本身只有一个大小不到 300KB 的文件，所以常作为嵌入式数据库内嵌在应用程序中。SQLite 生成的数据库文件是一个普通的磁盘文件，可以放在任何目录下。SQLite 是用 C 语言开发的，开放源代码，支持跨平台，最大支持 2048GB 数据，并且被所有的主流编程语言支持。可以说，SQLite 是一个非常优秀的嵌入式数据库。

SQLite 数据库的管理工具很多，比较常用的有 SQLite Expert Professional，其功能强大，几乎可以在可视化环境下完成所有的数据库操作。用户可以方便地使用它进行创建数据表和对数据记录进行增加、删除、修改、查询的操作。SQLite Expert Professional 的运行界面如图 8.1 所示。

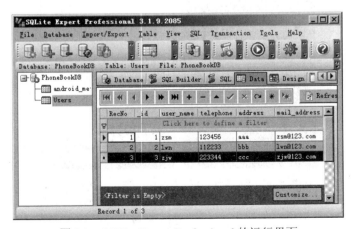

图 8.1　SQLite Expert Professional 的运行界面

在 Android 系统内部集成了 SQLite 数据库，所以，Android 应用程序可以很方便地使用 SQLite 数据库来存储数据。

在 Android SDK 的 tools 目录中有一个 SQLite 的命令行工具 sqlite3.exe。下面简单说明其使用方法和步骤：

（1）在控制台窗口的命令行中输入"adb shell"，进入 adb 调试的 shell 命令行状态。在"#"提示符下输入"cd/data/data/<packageName>/databases"（其中，<packageName>是包名，例如 com.ex08_01），进入项目存放数据库文件的目录中。

（2）输入"sqlite3<DatabaseName>"（其中，DatabaseName 是数据库名称，例如 sqlite3 phoneBookDB），进入 SQL 命令状态。

（3）在"sqlite>"提示符下输入 SQL 查询命令"select * from users"（其中，users 是数据库 phoneBookDB 中的数据表），显示数据表 users 中的全部记录。

（4）输入".quit"退出 SQL 命令状态，然后输入"exit"则退出 shell 命令行。

在命令行状态下执行 SQL 查询命令的过程如图 8.2 所示。

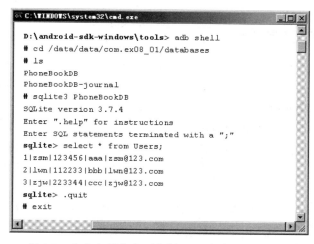

图 8.2　在命令行状态下执行 SQL 命令查询的过程

用户可以通过 DDMS 工具将数据库文件复制到本地计算机上，应用 SQLite Expert Professional 对数据库进行操作，完成后再通过 DDMS 工具放回到设备中。当需要操作大量数据时，这是比较方便的方法。

8.1.2　管理和操作 SQLite 数据库的对象

Android 提供了创建和使用 SQLite 数据库的 API（Application Programming Interface，应用程序编程接口）。在 Android 系统中，主要由类 SQLiteDatabase 和 SQLiteOpenHelper 对 SQLite 数据库进行管理和操作，下面分别对它们进行介绍。

1. SQLiteDatabase 类

在 Android 中，主要通过 SQLiteDatabase 对象对 SQLite 数据库进行管理。SQLiteDatabase 提供了一系列操作数据库的方法，可以对数据库进行创建、删除、执行 SQL 命令等操作，其常用方法见表 8-1。

表 8-1　SQLiteDatabase 的常用方法

方法	说明
openOrCreateDatabase(String path, SQLiteDatabase.CursorFactory factory)	打开或创建数据库
openDatabase(String path, SQLiteDatabase.CursorFactory factory, int flags)	打开指定的数据库
insert(String table, String nullColumnHack, ContentValues values)	新增一条记录
delete(String table, String whereClause, String[] whereArgs)	删除一条记录
query(String table, String[] columns, String selection, String[]selectionArgs, String groupBy, String having, String orderBy)	查询一条记录
update(String table, ContentValues values, String whereClause, String[] whereArgs)	修改记录
execSQL(String sql)	执行一条 SQL 语句
close()	关闭数据库

2. SQLiteOpenHelper 类

SQLiteOpenHelper 是 SQLiteDatabase 的一个辅助类。这个类主要用于创建数据库，并对数据库的版本进行管理。当在程序中调用这个类的 getWritableDatabase()方法或者 getReadableDatabase()方法的时候，如果数据库不存在，那么 Android 系统会自动创建一个数据库。

SQLiteOpenHelper 是一个抽象类，在使用时，一般要定义一个继承 SQLiteOpenHelper 的子类，并实现其方法。SQLiteOpenHelper 的方法见表 8-2。

表 8-2　SQLiteOpenHelper 的方法

方法	说明
onCreate(SQLiteDatabase)	首次生成数据库时调用该方法
onOpen(SQLiteDatabase)	调用已经打开的数据库
onUpgrade(SQLiteDatabase,int,int)	升级数据库时调用
getWritableDatabase()	以读/写方式创建或打开数据库
getReadableDatabase()	创建或打开数据库

8.1.3　SQLite 数据库的操作命令

对数据库的操作有 3 个层次，各层次的操作内容如下：
- 对数据库进行操作。建立数据库或删除数据库。
- 对数据表进行操作。建立、修改或删除数据库中的数据表。
- 对记录进行操作。对数据表中的数据记录进行添加、删除、修改、查询等操作。

下面按上述 3 个层次讲述 SQLite 数据库操作命令的使用方法。

1. 对数据库进行操作

1）创建数据库

创建数据库的方法有多种，可以应用 SQLiteDatabase 对象 openDatabase()方法及 openOrCreateDatabase()方法创建数据库，也可以应用 SQLiteOpenHelper 的子类创建数据库，还可以应用 Activity 继承于父类 android.content.Context 创建数据库的方法 openOrCreateDatabase()来创建数据库。

Context 类的 openOrCreateDatabase(name, mode, factory)方法有 3 个参数：

（1）第 1 个参数 name 为数据库名称；

（2）第 2 个参数 mode 为打开或创建数据库的模式，其模式有 3 种。

- MODE_PRIVATE：只可访问或调用模式，这是默认的模式。
- MODE_WORLD_READABLE：只读模式。
- MODE_WORLD_WRITEABLE：只写模式。

（3）第 3 个参数 factory 为查询数据的游标，通常为 null。

例如，要创建一个名为 PhoneBook.db 的数据库，其数据库的结构如下：

```
String TABLE_NAME="Users";          ← 数据表名
String ID="_id";                    ← ID
String USER_NAME="user_name";       ← 用户名
String ADDRESS="address";           ← 地址
String TELEPHONE="telephone";       ← 联系电话
String MAIL_ADDRESS="mail_address"; ← 电子邮箱
```

则用 Activity 的 openOrCreateDatabase()方法创建数据库的代码如下：

```
SQLiteDatabase  db;
String db_name="PhoneBook.db";
String sqlStr=
    "CREATE TABLE "+TABLE_NAME+" ("
        +ID+"INTEGER primary key autoincrement, "
        +USER_NAME+"text not null, "
        +TELEPHONE+"text not null, "
        +ADDRESS+"text not null, "
        +MAIL_ADDRESS+"text not null"+");";      ← 创建数据表的 SQL 语句
int mode=Context.MODE_PRIVATE;
db=this.openOrCreateDatabase(Database_name, mode, null);   ← 创建数据库
db.execSQL(sqlStr);      ← 执行创建数据库的 SQL 语句
```

这时，通过 DDMS 可以看到，在\data\data\xxxx（包名）\databases 下创建了数据库 PhoneBook.db。

2）删除数据库

当要删除一个指定的数据库文件时，需要应用 android.content.Context 类的 deleteDatabase (String name)方法来删除这个指定的数据库。

例如，要删除名为 PhoneBook.db 的数据库，可以使用如下代码：

```
            MainActivity.this.deleteDatabase("PhoneBook.db");
```
【例 8-1】 编写一个创建与删除数据库的演示程序。

（1）设计用户界面。在程序的用户界面中设置一个文本标签和两个按钮，如图 8.3 所示。

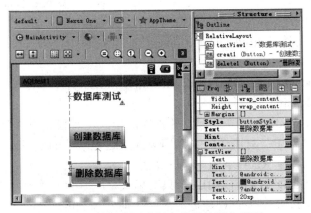

图 8.3 数据库测试程序的设计用户界面

（2）设计代码如下：

```
1   package com.ex08_01;
2   import android.os.Bundle;
3   import android.app.Activity;
4   import android.content.Context;
5   import android.database.sqlite.SQLiteDatabase;
6   import android.view.View;
7   import android.view.View.OnClickListener;
8   import android.widget.Button;
9   public class MainActivity extends Activity
10  {
11    Button creatBtn, deleteBtn;
12    @Override
13    public void onCreate(Bundle savedInstanceState)
14    {
15      super.onCreate(savedInstanceState);
16      setContentView(R.layout.activity_main);
17      creatBtn=(Button)findViewById(R.id.creat1);
18      creatBtn.setOnClickListener(new mClick());
19      deleteBtn=(Button)findViewById(R.id.delete1);
20      deleteBtn.setOnClickListener(new mClick());
21    }
22    class mClick implements OnClickListener
23    {
24      @Override
25      public void onClick(View arg0)
```

视频演示

```
26    {
27        if(arg0==creatBtn)        
28        {
29         DBCreate db=new DBCreate();        ← 实例化创建数据库类
30        }
31        else if(arg0==deleteBtn)
32        {
33         deleteDatabase(DBCreate.Database_name);   ← 删除数据库
34        }
35    }
36  }
37  class DBCreate
38  {
39    static final String Database_name="PhoneBook.db";   ← 数据库名
40    private DBCreate()
41    {
42       SQLiteDatabase db;
43       String TABLE_NAME="Users";          //数据表名
44       String ID="_id";                    //ID
45       String USER_NAME="user_name";       //用户名
46       String ADDRESS="address";           //地址            定义数据库结构
47       String TELEPHONE="telephone";       //联系电话
48       String MAIL_ADDRESS="mail_address"; //电子邮箱
49       String DATABASE_CREATE=
50           "CREATE TABLE "+TABLE_NAME+" ("
51           +ID+"INTEGER primary key autoincrement,"
52           +USER_NAME+"text not null, "              创建数据表
53           +TELEPHONE+"text not null, "              的 SQL 语句
54           +ADDRESS+"text not null, "
55           +MAIL_ADDRESS+"text not null"+");";
56       int mode=Context.MODE_PRIVATE;
57       db=openOrCreateDatabase(Database_name, mode, null);   ← 创建数据库
58       db.execSQL(DATABASE_CREATE);
59    }
60  }
61  }
```

运行程序,当单击"创建数据库"按钮后,通过 DDMS 工具调试监控视图,在\data\data\项目包名\databases 下,可以看到创建的数据库文件 PhoneBook.db,如图 8.4 所示。单击"删除数据库"按钮,数据库文件则被删除。

2. 对数据表进行操作

1) 创建数据表

创建数据表的步骤如下。

(1) 用 SQL 语句编写创建数据表的命令;

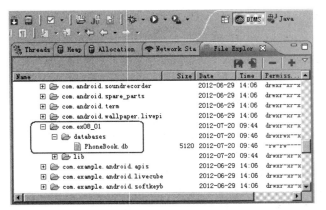

图 8.4 在 DDMS 视图中查看创建的数据库文件 PhoneBook.db

（2）调用 SQLiteDatabase 的 execSQL()方法执行 SQL 语句。

例如，在上面创建的数据库中，建立一个名为 Users 的数据表。

2）删除数据表

删除数据表的步骤与创建数据表类似，即先编写删除数据表的 SQL 语句,再调用 execSQL() 方法执行 SQL 语句。

例如，要删除名为 Users 的数据表，可执行如下代码：

```
String sql="DROP TABLE Users";    ← 删除数据表 Users 的 SQL 语句
db.execSQL(sql);    ← 执行 SQL 语句
```

3. 对数据记录进行操作

在数据表中，将数据表的列称为字段，数据表的每一行称为记录。对数据表中的数据进行操作处理，主要是对其记录进行操作处理。

对数据记录进行操作处理有两种方法。一种方法是编写一条对记录进行增、删、改、查的 SQL 语句，通过 exeSQL()方法来执行。另一种方法是使用 Android 系统的 SQLiteDatabase 对象的相应方法进行操作。对于使用 SQL 语句执行相应操作的方法，一般数据库书籍均有介绍，这里就不再赘述。下面介绍使用 SQLiteDatabase 对象操作数据记录的方法。

1）新增记录

新增记录使用 SQLiteDatabase 对象的 insert()方法实现。

insert(String table, String nullColumnHack, ContentValues values)方法中的 3 个参数的含义如下。

- 第 1 个参数 table：增加记录的数据表。
- 第 2 个参数 nullColumnHack：空列的默认值，通常为 null。
- 第 3 个参数 ContentValues：为 ContentValues 对象，其实就是一个键值对的字段名称，键名为表中的字段名，键值为要增加的记录数据值。通过 ContentValues 对象的 put()方法将数据存放到 ContentValues 对象中。

例如，下列代码分别把 4 个键值对数据存放到 values 对象中，其键名分别为 USER_NAME（用户名）、TELEPHONE（联系电话）、ADDRESS（住址）、MAIL_ADDRESS（电子邮箱）。

```
ContentValues values=new ContentValues();
values.put(USER_NAME, "张大山");
values.put(TELEPHONE, "13800012345");
values.put(ADDRESS, "南海仙人岛");
values.put(MAIL_ADDRESS, "abc@123.com");
db.insert("Users", null, values);
```
← 将数据存放到 values 中
← 将 values 数据添加到表 Users 中

2）修改记录

修改记录使用 SQLiteDatabase 对象的 update()方法。在 update(String table，ContentValues values，String whereClause，String[] whereArgs)方法中有 4 个参数，其含义如下。

- 第 1 个参数 table：修改记录的数据表。
- 第 2 个参数 ContentValues：ContentValues 对象，存放已修改数据的对象。
- 第 3 个参数 whereClause：修改数据的条件，相当于 SQL 语句的 where 子句。
- 第 4 个参数 whereArgs：修改数据值的数组。

例如，下列代码用用户名为张大山的经过修改的数据记录替换数据表中原来的数据记录，其他数据值不变：

```
ContentValues values=new ContentValues();
values.put(TELEPHONE, "13811011011");
values.put(ADDRESS, "东海彭湖湾");
String where=" USER_NAME = 张大山";
db.update("Users", values, where ,null);
```
← 将修改的数据存放到 values 中
← 用 values 数据替换表 Users 中满足条件的原数据

3）删除记录

删除记录使用 SQLiteDatabase 对象的 delete()方法。在 delete(String table,String whereClause, String[] whereArgs) 方法中有 3 个参数，其含义如下。

- 第 1 个参数 table：修改记录的数据表。
- 第 2 个参数 whereClause：删除数据的条件，相当于 SQL 语句的 where 子句。
- 第 3 个参数 whereArgs：删除条件的参数数组。

例如,下列代码将用户名为张大山的所有数据删除,即删除条件为"USER_NAME = 张大山"的记录：

```
String where="USER_NAME=张大山";
db.delete("Users", where ,null);
```
← 删除表 Users 中满足条件的记录

4）查询记录

在数据库的操作命令中，查询数据的命令是最丰富、最复杂的。在 SQLite 数据库中，使用 SQLiteDatabase 对象的 query()方法查询数据。query(String table, String[] columns, String selection, String[] selectionArgs, String groupBy, String having, String orderBy)方法有 7 个参数，其含义如下。

- 第 1 个参数 table：查询记录的数据表。
- 第 2 个参数 columns：查询的字段，如为 null，则为所有字段。
- 第 3 个参数 selection：查询条件，可以使用通配符 " ? "。
- 第 4 个参数 selectionArgs：参数数组，用于替换查询条件中的 " ? "。
- 第 5 个参数 groupBy：查询结果按指定字段分组。

- 第 6 个参数 having：限定分组的条件。
- 第 7 个参数 orderBy：查询结果的排序条件。

5）对查询结果 cursor 的处理

用 query()方法查询的数据均封装到查询结果 Cursor 对象中，Cursor 相当于 SQL 语句中 resultSet 结果集上的一个游标，可以向前、向后移动。Cursor 对象的常用方法如下。

- moveToFirst()：移动到第一行。
- moveToLast()：移动到最后一行。
- moveToNext()：向前移动一行。
- MoveToPrevious()：向后移动一行。
- moveToPosition(position)：移动到指定位置。
- isBeforeFirst()：判断是否指向第 1 条记录之前。
- isAfterLast()：判断是否指向最后一条记录之后。

例如，要查询用户表 Users 中的全部记录，并向前移动记录。代码如下：

```
Cursor cursor;
cursor=db.query("Users", null , null, null, null, null, null);
cursor.moveToNext();
```

【例 8-2】编写程序，建立一个通讯录，可以向前、向后浏览数据记录，也可以添加、修改、删除数据。

视频演示

（1）用户界面设计。在界面设计中，用一个垂直线性布局嵌套了 3 个水平线性布局和一个表格布局，从而把整个界面划分为 4 个部分。在第 1 个水平线性布局中嵌套了"建立数据库"和"打开数据库"按钮；在第 2 个水平线性布局中嵌套了"浏览通讯录"文本标签、"建立数据库"和"打开数据库"按钮；在第 3 个水平线性布局中嵌套了"添加""修改""删除"和"关闭通讯录"按钮；在表格布局中设置表格为 4 行 2 列，每一行均包含一个文本标签和一个文本编辑框。

用户界面布局设计如图 8.5 所示。

图 8.5 通讯录的界面布局设计

（2）数据库程序 DBConnection.java 的代码如下：

```
1   package com.ex08_02;
2   import android.content.Context;
3   import android.database.Cursor;
4   import android.database.sqlite.SQLiteDatabase;
5   import android.database.sqlite.SQLiteOpenHelper;
6
7   public class DBConnection extends SQLiteOpenHelper
8   {
9       static final String Database_name = "PhoneBook.db";    ← 定义数据库名
10      static final int Database_Version = 1;
11      SQLiteDatabase db;    ← 定义 SQLiteDatabase 数据库对象
12      public int id_this;
13      Cursor cursor;
14      //定义数据库名称及结构
15      static String TABLE_NAME="Users";    ← 数据表名
16      static String ID="_id";    ← ID
17      static String USER_NAME="user_name";    ← 用户名
18      static String ADDRESS="address";    ← 地址
19      static String TELEPHONE="telephone";    ← 联系电话
20      static String MAIL_ADDRESS="mail_address";    ← 电子邮箱
21      DBConnection(Context ctx)
22      {
23          super(ctx, Database_name, null, Database_Version);    ← 创建空数据库
24      }
25      public void onCreate(SQLiteDatabase database)
26      {
27          String sql="CREATE TABLE"+TABLE_NAME+"("
28              +ID+"INTEGER primary key autoincrement,"
29              +USER_NAME+"text not null,"
30              +TELEPHONE+"text not null,"
31              +ADDRESS+"text not null,"
32              +MAIL_ADDRESS+" text not null"+");";
33          database.execSQL(sql);
34      }
35      public void onUpgrade(SQLiteDatabase db, int oldVersion, int newVersion)
36      {          }
37  }
```

（创建数据表）

（3）控制程序 MainActivity.java 的代码如下：

```
1   package com.ex08_02;
2   import android.app.Activity;
3   import android.content.ContentValues;
4   import android.content.Context;
5   import android.database.Cursor;
```

```java
6   import android.database.sqlite.SQLiteDatabase;
7   import android.os.Bundle;
8   import android.view.View;
9   import android.view.View.OnClickListener;
10  import android.widget.Button;
11  import android.widget.EditText;
12
13  public class MainActivity extends Activity
14  {
15      static EditText mEditText01;
16      static EditText mEditText02;
17      static EditText mEditText03;
18      static EditText mEditText04;
19      Cursor cursor;
20      Button createBtn, openBtn, upBtn, downBtn;
21      Button addBtn, updateBtn, deleteBtn, closeBtn;
22      SQLiteDatabase db;
23      DBConnection helper;
24      public int id_this;
25      Bundle savedInstanceState;
26      //定义数据库名称及结构
27      static String TABLE_NAME = "Users";        // 数据表名
28      static String ID = "_id";                  // ID
29      static String USER_NAME = "user_name";     // 用户名
30      static String ADDRESS = "address";         // 地址
31      static String TELEPHONE = "telephone";     // 联系电话
32      static String MAIL_ADDRESS = "mail_address"; // 电子邮箱
33      @Override
34      public void onCreate(Bundle savedInstanceState)
35      {
36          super.onCreate(savedInstanceState);
37          setContentView(R.layout.main);
38          mEditText01=(EditText)findViewById(R.id.EditText01);
39          mEditText02=(EditText)findViewById(R.id.EditText02);   // 初始化文
40          mEditText03=(EditText)findViewById(R.id.EditText03);   // 本编辑框
41          mEditText04=(EditText)findViewById(R.id.EditText04);
42          createBtn=(Button)findViewById(R.id.createDatabase1);
43          createBtn.setOnClickListener(new ClickEvent());
44          openBtn=(Button)findViewById(R.id.openDatabase1);
45          openBtn.setOnClickListener(new ClickEvent());
46          upBtn=(Button)findViewById(R.id.up1);                  // 初始化
47          upBtn.setOnClickListener(new ClickEvent());            // 按钮
48          downBtn=(Button)findViewById(R.id.down1);
49          downBtn.setOnClickListener(new ClickEvent());
50          addBtn=(Button)findViewById(R.id.add1);
```

```
51      addBtn.setOnClickListener(new ClickEvent());
52      updateBtn=(Button)findViewById(R.id.update1);
53      updateBtn.setOnClickListener(new ClickEvent());
54      deleteBtn=(Button)findViewById(R.id.delete1);
55      deleteBtn.setOnClickListener(new ClickEvent());
56      closeBtn=(Button)findViewById(R.id.clear1);
57      closeBtn.setOnClickListener(new ClickEvent());
58  }
59  class ClickEvent implements OnClickListener
60  {
61      public void onClick(View v)
62      {
63          switch(v.getId())
64          {
65          case R.id.createDatabase1:
66              helper=new DBConnection(MainActivity.this);
67              SQLiteDatabase db=helper.getWritableDatabase();
68              break;
69          case R.id.openDatabase1:
70              db=openOrCreateDatabase("PhoneBook.db",
71                      Context.MODE_PRIVATE, null) ;
72              cursor=db.query("Users",
73                  null , null, null, null, null, null);
74              cursor.moveToNext();
75              upBtn.setClickable(true);
76              downBtn.setClickable(true);
77              deleteBtn.setClickable(true);
78              updateBtn.setClickable(true);
79              break;
80          case  R.id.up1:
81              if(!cursor.isFirst())
82                  cursor.moveToPrevious();
83              datashow();
84              break;
85          case R.id.down1:
86              if(!cursor.isLast())
87                  cursor.moveToNext();
88              datashow();
89              break;
90          case R.id.add1:
91              add();
92              onCreate(savedInstanceState);
93              break;
```

```
94          case R.id.update1:
95             update();
96             onCreate(savedInstanceState);     ← 单击"修改"按钮,更新一行数据
97             break;
98          case R.id.delete1:
99             delete();
100            onCreate(savedInstanceState);     ← 单击"删除"按钮,删除一行数据
101            break;
102         case R.id.clear1:
103            cursor.close();                    ← 单击"关闭通讯录"按钮,关闭数据库
104            mEditText01.setText("数据库已关闭");
105            mEditText02.setText("数据库已关闭");
106            mEditText03.setText("数据库已关闭");
107            mEditText04.setText("数据库已关闭");
108            upBtn.setClickable(false);
109            downBtn.setClickable(false);       ← 使按钮不可用
110            deleteBtn.setClickable(false);
111            updateBtn.setClickable(false);
112            break;
113         }
114      }
115   }
116   /*显示记录*/
117   void datashow()
118   {
119    id_this=Integer.parseInt(cursor.getString(0));
120    String user_name_this=cursor.getString(1);
121    String telephone_this=cursor.getString(2);     ← 读取各字段数据
122    String address_this=cursor.getString(3);
123    String mail_address_this=cursor.getString(4);
124    mEditText01.setText(user_name_this);
125    mEditText02.setText(telephone_this);
126    mEditText03.setText(address_this);
127    mEditText04.setText(mail_address_this);
128   }
129   /*添加记录*/
130   void add()
131   {
132   ContentValues values1=new ContentValues();
133   values1.put(USER_NAME, MainActivity.mEditText01.getText().toString());
134   values1.put(TELEPHONE, MainActivity.mEditText02.getText().toString());
135   values1.put(ADDRESS, MainActivity.mEditText03.getText().toString());
136   values1.put(MAIL_ADDRESS,
137               MainActivity.mEditText04.getText().toString());
138   SQLiteDatabase db2=helper.getWritableDatabase();
```

```
139     db2.insert(TABLE_NAME, null, values1);        ← 将数据插入数据表
140     db2.close();
141   }
142   /*修改记录*/
143   void update()
144   {
145     ContentValues values=new ContentValues();
146     values.put(USER_NAME, MainActivity.mEditText01.getText().toString());
147     values.put(TELEPHONE, MainActivity.mEditText02.getText().toString());
148     values.put(ADDRESS, MainActivity.mEditText03.getText().toString());
149     values.put(MAIL_ADDRESS,
150                      MainActivity.mEditText04.getText().toString());
151     String where1=ID+"="+id_this;
152     SQLiteDatabase db1=helper.getWritableDatabase();
153     db1.update(TABLE_NAME, values, where1 ,null);  ← 替换数据表中的原数据
154     db1.close();
155   }
156   /*删除记录*/
157   void delete()
158   {
159     String where=ID+"="+id_this;
160     db.delete(TABLE_NAME, where ,null);            ← 删除数据表中的数据
161     db=helper.getWritableDatabase();
162     db.close();
163   }
164 }
```

程序的运行结果如图 8.6 所示。

图 8.6　数据库运行示例

8.2 文件处理

8.2.1 输入流和输出流

程序可以理解为数据输入、输出及处理的过程，在程序执行过程中，通常需要读取处理数据，并且将处理后的结果保存起来。Android 系统提供了对数据流进行输入、输出的方法。

1. 流的概念

流是比文件所包含范围更广的概念。流是一个可被顺序访问的数据序列，是对计算机输入数据和输出数据的抽象。

就流的运行方向来说，流分为输入流和输出流。输入流将外部数据输送到微处理器 CPU 中，例如从 SD 卡中读取信息、从网络上读取信息等；输出流将数据发送到外部设备，例如向 SD 卡保存文件、向网络中发送信息或在屏幕上显示文件内容等。因此，可以将"流"看作数据从一种设备流向另一种设备的过程（如图 8.7 所示）。它最大的特点就是数据的获取和发送均按数据序列顺序进行：每一个数据都必须等待排在它前面的数据读入或送出之后才能被读/写，而不能够随意选择输入、输出的位置。

图 8.7 "流"是数据从一种设备流向另一种设备的过程

2. 文件与目录管理的 File 类

Android 系统处理文件时直接调用 Java 语言的 java.io 包中的 File 类。每个 File 类的对象都对应了系统的一个文件或目录，所以创建 File 类对象时需指明它所对应的文件或目录名。为了便于建立 File 对象，File 类共提供了 3 个不同的构造函数，以不同的参数形式灵活地接收文件和目录名信息。

1）File（String path）

这个构造方法的字符串参数 path 指明了新创建的 File 对象对应的文件及其路径。下面是用该构造方法创建 File 对象的例子：

```
File f1=new File("\data\data\jtest");
File f2=new File("testfile.dat");
```

2）File（String path, String name）

这个构造方法有两个参数，path 表示所对应的文件或目录的绝对或相对路径，name 表示文件或目录名。将路径与名称分开的好处是相同路径的文件或目录可共享同一个路径字符串，这样管理和修改都比较方便。例如：

```
File f4=new File(" \sdcard", "file.dat");
```

3）File（File dir, String name）

这个构造方法使用另一个已有的某 SD 卡目录的 File 对象作为第一个参数，表示文件或目录的路径，第二个字符串参数表示文件或目录名。例如：

```
String  sdir="data"+System.dirSep+"jtest";
String  sfile="FileIO.data";
File  Fdir=new File(sdir);
File  Ffile=new File(Fdir, sfile);
```

其中，System.dirSep 为当前系统的目录分隔符（"/"）。

一个对应于某文件或目录的 File 对象一经创建，即可通过调用它的方法来获得文件或目录的属性。建立文件类的一个实例后，可以查询这个文件对象，用测试方法获得有关文件或目录的有关信息，如检测文件和目录的属性。File 类的常用方法见表 8-3。

表 8-3 File 类的常用方法

方　　法	说　　明
exists()	判断文件或目录是否存在
isFile()	判断对象是否是文件
isDirectory()	判断对象是否是目录
getName()	返回文件名或目录名
getPath()	返回文件或目录的路径
length()	返回文件的字节数
renameTo(File newFile)	将文件重命名成 newFile 对应的文件名
delete()	将当前文件删除
mkdir()	创建当前目录的子目录

3. 文件输入流和输出流

在 Android 中，处理二进制文件使用字节输入和输出流，处理字符文件使用字符输入和输出流。对文件进行输入和输出处理的流有 4 个类。

- FileInputStream：字节文件输入流。
- FileOutputStream：字节文件输出流。
- FileReader：字符文件输入流。
- FileWriter：字符文件输出流。

8.2.2 处理文件流

1. 文件输出流保存文件

1）FileOutputStream 类

FileOutputStream 类是从 OutputStream 类派生出来的输出类，它具有向文件中写数据的能力。它的构造方法有以下 3 种形式：

- FileOutputStream(String filename);
- FileOutputStream(File file);
- FileOutputStream(FileDescriptor fdObj);

其中各参数的含义如下。

- String filename：指定的文件名，包括路径。
- File file：指定的文件对象。
- FileDescriptor fdObj：指定的文件描述符。

用户也可以通过 Context.openFileOutput()方法获取 FileOutputStream 对象。

2）把字节发送到文件输出流的 write()方法

输出流只是建立了一条通往数据要去的目的地的通道，数据并不会自动进入输出流通道，因此要使用文件输出流的 write()方法把字节发送到输出流。

使用 write()方法有 3 种形式：

- write(int b);
- write(byte[ ] b);
- write(byte[ ] b, int off, int len);

其中各参数的含义如下。

- write(int b)：将指定字节写入此文件输出流。
- write(byte[] b)：将 b.length 个字节从指定字节数组写入此文件输出流中。
- write(byte[] b, int off, int len)：将指定字节数组中从偏移量 off 开始的 len 个字节写入此文件输出流。

【例 8-3】把字符串"Hello World!"保存到本地资源的 test.txt 文件中。

视频演示

在项目设计中，设置一个"保存文件"按钮，其按钮事件调用下列方法：

```
1  void savefile()
2  {
3      String fileName="test.txt";
4      String str="Hello World!";
5      FileOutputStream f_out;
6      try {
7          f_out=openFileOutput(fileName, Context.MODE_PRIVATE);   ←文件输出流
8          f_out.write(str.getBytes());   ←写操作
9      }
10     catch(FileNotFoundException e) {e.printStackTrace();}
11     catch(IOException e) {e.printStackTrace();}
12 }
```

则文件 test.txt 保存在\data\data\<包名>\files\目录之下。使用 DDMS 工具可以查看保存在本地资源目录下的文件，如图 8.8 所示。

2. 文件输入流读取文件

1）FileInputStream 类

FileInputStream 类是从 InputStream 类中派生出来的输入流类，用于处理二进制文件的

输入操作。其构造方法有下面 3 种形式：

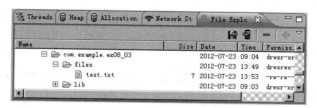

图 8.8 使用 DDMS 工具查看保存到本地资源目录下的文件

- FileInputStream(String filename);
- FileInputStream(File file);
- FileInputStream(FileDescriptor fdObj);

其参数的含义和 FileInputStream 一样。

用户也可以通过 Context.openFileInput()方法获取 FileInputStream 对象。

2）从文件输入流中读取字节的 read()方法

文件输入流只是建立了一条通往数据的通道，应用程序可以通过这个通道读取数据，要实现读取数据的操作，需要使用 read()方法。

使用 read()方法有 3 种形式：

- int read();
- int read(byte b[ ]);
- int read(byte b[ ],int off, int len);

第 1 种形式每次只能从输入流中读取一个字节的数据。该方法返回的是 0～255 的一个整数值，若为文本类型的数据则返回的是 ASCII 值。如果该方法到达输入流的末尾，则返回-1。

第 2 种形式和第 3 种形式以字节型数组作为参数，一次可以读取多个字节，读入的字节数据直接放入字节数组 b 中，并返回实际读取的字节个数。如果该方法到达输入流的末尾，则返回-1。

第 3 种形式设置了偏移量（off）。这里的偏移量是指可以从字节型数组的 off 位置起，读取 len 个数据。

视频演示

【例 8-4】读取本地资源文件 test.txt 中的内容。

在项目设计中，设置一个"读取资源文件"按钮，其按钮事件调用下列方法：

```
1   void readfile()
2   {
3       String fileName="test.txt", str;
4       byte[]  buffer=new byte[1024];        ← 设置一个字节数组，用于存放读取的数据
5       FileInputStream in_file=null;
6       try {
7           in_file=openFileInput(fileName);   ← 文件输入流
8           int bytes=in_file.read(buffer);
9           str=new String(buffer, 0, bytes);  ← 将读取到的文件数据转换成字符串
10          Toast.makeText(MainActivity.this,
11               "文件内容: " + str, Toast.LENGTH_LONG).show();
```

```
12      }
13      catch(FileNotFoundException e) { System.out.print("文件不存在");}
14      catch(IOException e) { System.out.print("IO流错误");
15  }
```

3. 对 SD 卡文件的读/写

上述应用文件流对文件的读/写操作也适用于 SD 卡，但在处理上稍有不同，因为这里要考虑对 SD 卡的读/写权限。

1）环境变量访问类 Environment

Environment 为提供对环境变量访问类，在 Android 程序中对 SD 卡文件进行读/写操作时，经常需要应用它的以下两个方法。

- getExternalStorageState()：获取当前存储设备的状态。
- getExternalStorageDirectory()：获取 SD 卡的根目录。

2）读/写 SD 卡的权限

Environment 的主要常量为 Environment.MEDIA_MOUNTED，表示对 SD 卡具有读/写权限。判断是否具有对 SD 卡文件进行读/写操作的权限通常使用下列条件语句：

```
if(Environment.getExternalStorageState().equals(Environment.MEDIA_MOUNTED))
```

在 AndroidManifest.xml 文件中，要添加允许对 SD 卡进行操作的权限语句。

- 允许在 SD 卡中创建及删除文件的权限语句：

```
<uses-permission
   android:name="android.permission.MOUNT_UNMOUNT_FILESYSTEMS">
</uses-permission>
```

- 允许向 SD 卡中写入数据的权限语句：

```
<uses-permission
   android:name="android.permission.WRITE_EXTERNAL_STORAGE">
</uses-permission>
```

【例 8-5】读取与保存文件的应用程序示例。

新建应用项目，设置一个文本输入框和 4 个按钮，按钮分别为"保存文件""保存到 SD 卡""读取资源文件""读取 SD 卡文件"。

在 AndroidManifest.xml 文件中，添加允许对 SD 卡进行创建文件和写入数据的权限语句。

视频演示

编写控制程序 MainActivity.java，代码如下：

```
1   package com.example.ex08_05;
2   import java.io.File;
3   import java.io.FileInputStream;
4   import java.io.FileNotFoundException;
5   import java.io.FileOutputStream;
6   import java.io.IOException;
7   import android.os.Bundle;
8   import android.os.Environment;
9   import android.view.View;
10  import android.view.View.OnClickListener;
```

```
11  import android.widget.Button;
12  import android.widget.EditText;
13  import android.widget.Toast;
14  import android.app.Activity;
15  import android.content.Context;
16  public class MainActivity extends Activity
17  {
18    Button saveBtn, readBtn, savesdBtn, readsdBtn;
19    EditText edit;
20    String fileName="test.txt";
21    String str;
22    @Override
23    public void onCreate(Bundle savedInstanceState)
24    {
25        super.onCreate(savedInstanceState);
26        setContentView(R.layout.activity_main);
27        edit=(EditText)findViewById(R.id.edit1);
28        saveBtn=(Button)findViewById(R.id.button1);
29        saveBtn.setOnClickListener(new mClick());
30        readBtn=(Button)findViewById(R.id.button2);
31        readBtn.setOnClickListener(new mClick());
32        savesdBtn=(Button)findViewById(R.id.button3);
33        savesdBtn.setOnClickListener(new mClick());
34        readsdBtn=(Button)findViewById(R.id.button4);
35        readsdBtn.setOnClickListener(new mClick());
36    }
37    //按钮事件
38    class mClick implements OnClickListener
39    {
40      @Override
41      public void onClick(View arg0)
42      {
43          if(arg0==saveBtn)
44      {
45          savefile();       ← 写入文件
46      }
47      else if(arg0==readBtn)
48      {
49          readfile(fileName);   ← 读取资源文件数据
50      }
51      else if(arg0==savesdBtn)
52      {
53          setRegist();   //动态获取本地存储卡（SD 卡）读写权限
54          saveSDcar();    ← 写入到 SD 卡
55      }
56      else if(arg0==readsdBtn)
57      {
58          readsdcard(fileName);   ← 读取 SD 卡文件数据
59      }
60     }
61  }
62
63  void savefile()
64  {
65    str= edit.getText().toString();
66    try {
```

```
67            FileOutputStream f_out=                    ← 创建文件输出流
68            openFileOutput(fileName, Context.MODE_PRIVATE);
69            f_out.write(str.getBytes());           ← 文件输出流按字节输出数据到文件中
70        }catch (FileNotFoundException e) {
71            e.printStackTrace();
72        } catch (IOException e) {
73            e.printStackTrace();
74        }
75    }
76    //读取文件数据
77    void readfile(String fileName)
78    {
79        byte[] buffer=new byte[1024];              ← 创建字节数组存放数据
80        FileInputStream in_file=null;
81    try {
82            in_file=openFileInput(fileName);       ← 文件输入流将读到的数据放入字
83            int bytes=in_file.read(buffer);            节数组，再将字节转换成字符串
84            str=new String(buffer, 0, bytes);
85            Toast.makeText(MainActivity.this,
86                    "文件内容: " + str, Toast.LENGTH_LONG).show();
87    } catch (FileNotFoundException e) { System.out.print("文件不存在");}
88      catch (IOException e) { System.out.print("IO流错误");}
89    }
90    //保存文件到SD卡
91    void saveSDcar()
92    {
93     str= edit.getText().toString();
94     if(Environment.getExternalStorageState()       ← 判断SD卡是否允许读/写操作
95           .equals(Environment.MEDIA_MOUNTED))
96    {
97      File path = Environment.getExternalStorageDirectory();  ← 获取SD卡
98      File sdfile = new File(path, fileName);                    目录路径
99      try {
100         FileOutputStream f_out=new FileOutputStream(sdfile);
101         f_out.write(str.getBytes());           ← 文件输出流将数据写入SD卡
102         Toast.makeText(MainActivity.this,
103                 "文件保存到SD卡", Toast.LENGTH_LONG).show();
104        }catch (FileNotFoundException e)
105        {
106            e.printStackTrace();
107        } catch (IOException e) {
108            e.printStackTrace();
109        }
110   }
111   }
112   //从SD卡读取文件内容
113   void readsdcard(String fileName)
114   {
115     if(Environment.getExternalStorageState()      ← 判断SD卡是否允许读/写操作
116           .equals(Environment.MEDIA_MOUNTED))
117    {
118        File path=Environment
119                  .getExternalStorageDirectory();  ← 获取SD卡目录路径
120        File sdfile=new File(path, fileName);
121        try {
```

```
122                FileInputStream in_file=new  FileInputStream(sdfile);
123                byte[]  buffer=new byte[1024];
124                int bytes=in_file.read(buffer);      ←── 读取文件数据到字节数组
125                str=new String(buffer, 0, bytes);
126                Toast.makeText(MainActivity.this,
127                    "文件内容: " + str, Toast.LENGTH_LONG).show();
128          } catch(FileNotFoundException e){ System.out.print("文件不存在");}
129            catch(IOException e)  { System.out.print("IO流错误");}
130      }
131 }
132 private void setRegist() {
          /*
           * 动态获取权限
           * Android 6.0 新特性，一些保护权限，如，文件读写除了要在
           * AndroidManifest 中声明权限，还要使用如下代码动态获取
           */
133    if (Build.VERSION.SDK_INT >= 23) {//大于23是指Android 6.0以后版本
134    int REQUEST_CODE_CONTACT = 101;
135    String[] PERMISSIONS_STORAGE = {
136              Manifest.permission.READ_EXTERNAL_STORAGE,
137              Manifest.permission.WRITE_EXTERNAL_STORAGE};
       //验证是否许可权限
138    for (String str : PERMISSIONS_STORAGE) {
139    if (this.checkSelfPermission(str) !=
140              PackageManager.PERMISSION_GRANTED ) {
       //申请权限
141    this.requestPermissions(PERMISSIONS_STORAGE,REQUEST_CODE_CONTACT);
142    return;
143          }
144       }
145    }
146 }
```

程序的运行结果如图 8.9 所示。

图 8.9　读写文件示例

8.3 轻量级存储 SharedPreferences

Android 系统提供了存储少量数据的轻量级的数据存储方式 SharedPreferences。该存储方式类似于 Web 程序中的 Cookie，通常用它来保存一些配置文件数据、用户名及密码等。SharedPreferences 采用"键名-键值"的键值对形式组织和管理数据，其数据存储在 XML 格式文件中。

使用 SharedPreferences 方式存储数据需要用到 SharedPreferences 和 SharedPreferences.Editor 接口，这两个接口在 android.content 包中。

SharedPreferences 接口由 Context.getSharedPreferences (String name, int mode)方法构造，它有两个参数。

第 1 个参数 name 为保存数据的文件名，因为 SharedPreferences 是使用 XML 文件保存数据，getSharedPreferences(name,mode)方法的第 1 个参数用于指定该文件的名称，不用带扩展名，扩展名会由 Android 自动加上。该 XML 文件存放在\data\data\<包名>\shared_prefs 目录下。

第 2 个参数 mode 为操作模式，它有以下 4 个取值。

- MODE_PRIVATE：默认形式，配置文件只允许本程序和享有本程序 ID 的程序访问。
- MODE_WORLD_READABLE：允许其他应用程序读文件。
- MODE_WORLD_WRITEABLE：允许其他应用程序写文件。
- MODE_MULTI_PROCESS：主要用于多任务，当多个进程共同访问的时候，必须指定这个标签。

SharedPreferences 接口的常用方法见表 8-4。

表 8-4 SharedPreferences 接口的常用方法

方 法	说 明
edit()	建立一个 SharedPreferences.Editor 对象
contains(String key)	判断是否包含该键值
getAll()	返回所有配置信息
getBoolean(String key,Boolean defValue)	获得一个 Boolean 类型数据
getFloat(String key,float defValue)	获得一个 Float 类型数据
getInt(String key,int defValue)	获得一个 Int 类型数据
getLong(String key,long defValue)	获得一个 Long 类型数据
getString(String key,String defValue)	获得一个 String 类型数据

SharedPreferences.Editor 接口用于存储 SharedPreference 对象的数据值，SharedPreferences.Editor 接口的常用方法见表 8-5。

表 8-5 SharedPreferences.Editor 接口的常用方法

方 法	说 明
clear()	清除所有数据值
commit()	保存数据
putBoolean(String key, boolean value)	保存一个 Boolean 类型数据

续表

方　　法	说　　明
putFloat(String key, float value)	保存一个 Float 类型数据
putInt(String key, int value)	保存一个 Int 类型数据
putLong(String key, long value)	保存一个 Long 类型数据
putString(String key, String value)	保存一个 String 类型数据
remove(String key)	删除键名 key 所对应的数据值

SharedPreference.Editor 的 put×××方法以键值对的形式存储数据，最后一定要调用 commit 方法提交数据，这样文件才能保存。

读取数据非常简单，直接调用 SharedPreference 对象相应的 get×××方法即可获得数据。

【例 8-6】应用 SharedPreferences 对象将一个客户的联系电话保存到电话簿中。

设客户名为 zsm，其电话为 123456，电话簿的文件名为 phoneBook，其数据的"键名-键值"对为（"name", "zsm"）和（"phone", "123456"）。

代码如下：

视频演示

```
1    package com.example.ex08_06;
2    import android.os.Bundle;
3    import android.app.Activity;
4    import android.content.Context;
5    import android.content.SharedPreferences;
6    import android.view.View;
7    import android.view.View.OnClickListener;
8    import android.widget.Button;
9    import android.widget.Toast;
10
11   public class MainActivity extends Activity
12   {
13       SharedPreferences settings;
14       Button saveBtn;
15       @Override
16       public void onCreate(Bundle savedInstanceState)
17       {
18           super.onCreate(savedInstanceState);
19           setContentView(R.layout.activity_main);
20           saveBtn=(Button)findViewById(R.id.button1);
21           saveBtn.setOnClickListener(new mClick());
22       }
23       //按钮事件
24       class mClick implements OnClickListener
25       {
26           public void onClick(View arg0)
27           {
28               settings = getSharedPreferences("phoneBook", Context.MODE_PRIVATE);
```

```
29          SharedPreferences.Editor editor=settings.edit();
30          editor.putString("name", "zsm");
31          editor.putString("phone", "123456");
32          editor.commit();
33          Toast.makeText(MainActivity.this, "保存成功! ", Toast.LENGTH_LONG);
34      }
35  }
36 }
```

数据文件 phoneBook.xml 保存在\data\data\com.example.ex08_06（包名）\shared_prefs 之下，扩展名.xml 由系统自动生成。应用 DDMS 工具可以查看到该文件，如图 8.10 所示。

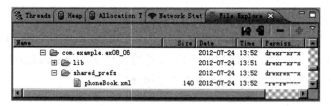

图 8.10　使用 DDMS 工具查看保存在\data\data\<包名>\shared_prefs 下的 xml 文件

单击 DDMS 工具中的 按钮，可以将数据文件 phoneBook.xml 从模拟器中保存到计算机的磁盘中。打开该 XML 格式的数据文件 phoneBook.xml，其内容如下：

```
<?xml version='1.0' encoding='utf-8' standalone='yes' ?>
<map>
<string name="phone">123456</string>
<string name="name">zsm</string>
</map>
```

习　题　8

1．编写一个小型商场的销售管理系统，使之可以输入商品的名称、数量、单价，并具有汇总功能。

2．编写一个如图 8.11 所示的简易记事本，以文件形式保存。

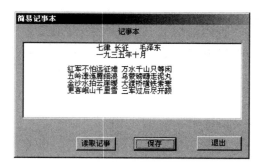

图 8.11　简易记事本

第 9 章　网　络　通　信

网络应用的核心思想是使连入网络的不同计算机能够跨越空间协同工作,这首先要求它们之间能够准确、迅速地传递信息。Java 是一门非常适合于分布计算环境的语言,网络应用是它的重要应用之一,尤其是它具有非常好的 Internet 网络程序设计功能。Java 的这种特性来源于它独有的一套用于网络的 API,这些 API 是一系列的类和接口,均位于 java.net 和 javax.net 包中。

本章将介绍 Android 用于编写网络通信程序的一些实例,其中重点介绍客户机/服务器的应用程序及 Web 视图应用程序的设计方法。

9.1　网络编程的基础知识

9.1.1　IP 地址和端口号

1. IP 地址

网络中连接了很多计算机,假设计算机 A 向计算机 B 发送信息,若网络中还有第三台计算机 C,那么主机 A 怎么知道信息被正确地传送到主机 B 而不是被传送到主机 C 中了呢?如图 9.1 所示。

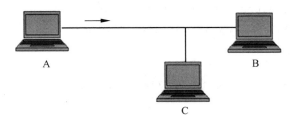

图 9.1　主机 A 向主机 B 发送信息

网络中的每台计算机都必须有一个唯一的 IP 地址作为标识,该地址通常写做一组由"."号分隔的十进制数,例如,思维论坛的服务器地址为 218.5.77.187。IP 地址均由 4 个部分组成,每个部分的范围都是 0~255。

值得注意的是,IP 地址都是 32 位地址,这是 IP 协议版本 4（简称 IPv4）规定的,目前由于 IPv4 地址已近耗尽,所以 IPv6 地址正逐渐代替 IPv4 地址,IPv6 地址是 128 位无符号整数。

在 Java.net 包中,IP 地址由一个称为 InetAddress 的特殊类来描述。这个类提供了 3 个用来获得一个 InetAddress 类的实例的静态方法。

（1）getLocalHost()：返回一个本地主机的 IP 地址。

（2）getByName(String host)：返回对应指定主机的 IP 地址。

（3）getAllByName(String host)：对于某个主机有多个 IP 地址（多宿主机），可用于得到一个 IP 地址数组。

此外，对于一个 InetAddress 的实例可以使用 getAddress()获得一个用字节数组形式表示的 IP 地址，可以使用 getHostName()做反向查询，获得对应某个 IP 地址的主机名。

【例 9-1】 通过域名查找 IP 地址。

```
1   package com.example.iptest;
2   import java.net.InetAddress;
3   import java.net.UnknownHostException;
4   import android.os.Bundle;
5   import android.view.View;
6   import android.view.View.OnClickListener;
7   import android.widget.Button;
8   import android.widget.Toast;
9   import android.app.Activity;
10  
11  public class MainActivity extends Activity
12  {
13    Button IPBtn;
14    @Override
15    public void onCreate(Bundle savedInstanceState)
16    {
17        super.onCreate(savedInstanceState);
18        setContentView(R.layout.activity_main);
19        IPBtn=(Button)findViewById(R.id.button1);
20        IPBtn.setOnClickListener(new mClick());
21    }
22  
23    class mClick implements OnClickListener
24    {
25     @Override
26     public void onClick(View arg0)
27     {
28      String str;
29      try{
30         InetAddress zsm_address=InetAddress.getByName("www.zsm8.com");
31          str="思维论坛的 IP 地址为: \n"+zsm_address.toString();
32       }
33      catch(UnknownHostException e)
34        {
```

视频演示

```
35                    str="无法找到思维论坛";
36                }
37           Toast.makeText(MainActivity.this, str, Toast.LENGTH_LONG).show();
38        }
39    }
40 }
```

网络程序需要在配置文件 AndroidManifest.xml 中添加允许访问网络的权限语句：

`<uses-permission android:name="android.permission.INTERNET" />`

程序的运行结果如图 9.2 所示。

图 9.2 通过域名查找 IP 地址

在上面的例子中将第 30 行的 getByName()方法改为 getLocalHost()方法，则显示设备的 IP 地址。

2. 端口

由于一台计算机上可同时运行多个网络程序，IP 地址只能保证把数据信息送到该计算机，但无法知道要把这些数据交给该主机上的哪个网络程序，因此，用"端口号"来标识正在计算机上运行的进程（程序）。每个被发送的网络数据包也都包含"端口号"，用于将该数据帧交给具有相同端口号的应用程序来处理。

例如，在一个网络程序中指定了所用的端口号为 52000，那么其他网络程序（例如端口号为 13）发送给这个网络程序的数据包必须包含 52000 端口号。当数据到达计算机后，驱动程序根据数据包中的端口号即可知道要将这个数据包交给哪个网络程序，如图 9.3 所示。

端口号是一个整数，其取值范围为 0~65535。同一台计算机上不能同时运行两个有相同端口号的进程。通常，0~1023 的端口号作为保留端口号，用于一些网络系统服务和应用，用户的普通网络应用程序应该使用 1024 之后的端口号，从而避免端口号冲突。

3. TCP 与 UDP

在网络协议中，有两个高级协议是网络应用程序中常用的，它们是传输控制协议（Transmission Control Protocol，TCP）和用户数据报协议（User Datagram Protocol，UDP）。

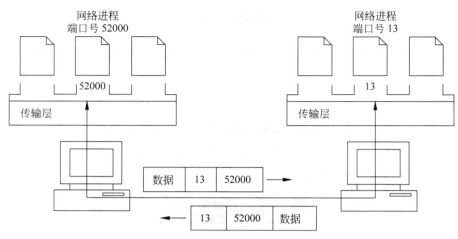

图 9.3 用"端口号"来标识进程

　　TCP 是面向连接的通信协议，提供两台计算机之间的可靠无差错的数据传输。应用程序利用 TCP 进行通信时，信息源与信息目标之间会建立一个虚连接。这个连接一旦建立成功，两台计算机之间就可以把数据当作一个双向字节流进行交换。接收方对于接收到的每一个数据包都会发送一个确认信息，发送方只有在收到接收方的确认信息后才发送下一个数据包，通过这种确认机制保证数据传输无差错。

　　UDP 是无连接通信协议，UDP 不保证可靠数据的传输。简单地说，如果一个主机向另外一台主机发送数据，这一数据就会立即发送，而不管另外一台主机是否已准备接收数据。如果另外一台主机收到了数据，它不会确认收到与否。这一过程，类似于从邮局发送信件，我们无法确定收信人一定收到了发出去的信件。

9.1.2 套接字

1. 什么是套接字

　　大家已经知道，通过 IP 地址可以在网络上找到主机，通过端口可以找到主机上正在运行的网络程序。在 TCP/IP 通信协议中，套接字（Socket）就是 IP 地址与端口号的组合。如图 9.4 所示，IP 地址 193.14.26.7 与端口号 13 组成一个套接字。

　　Java 使用了 TCP/IP 套接字机制，并使用一些类来实现套接字中的概念。Java 中的套接字提供了在一台处理机上执行的应用程序与在另一台处理机上执行的应用程序之间进行连接的功能。

　　网络通信，准确地说，不能仅说成两台计算机之间在通信，而是两台计算机上执行的网络应用程序（进程）之间在收发数据。

　　当两个网络程序需要通信时，它们可以通过使用 Socket 类建立套接字连接。可以把套接字连接想象为一个电话呼叫，当呼叫完成后，通话的任何一方都可以随时讲话。但是在最初建立呼叫时，必须有一方主动呼叫，而另一方正在监听铃声。这时，把呼叫方称为"客户端"，把负责监听的一方称为"服务器端"。

2. 客户端建立套接字 Socket 对象

　　在客户端使用 Socket 类建立向指定服务器 IP 和端口号连接的套接字，其构造方法如下：

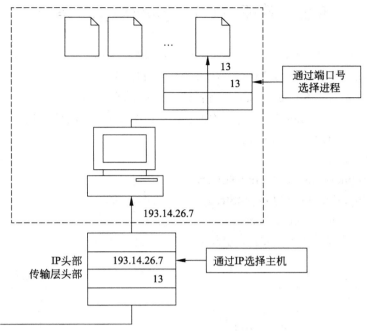

图 9.4 套接字是 IP 地址和端口号组合

```
Socket(host_IP, prot);
```

其中，host_IP 是服务器的 IP 地址，prot 是一个端口号。

由于建立 Socket 对象可能发生 IOException 异常，因此，在建立 Socket 对象时要使用 try-cahch 结构处理异常事件。

Socket 有下列两种主要方法。

（1）getInputStream()：获得一个输入流，读取从网络线路上传送来的数据信息。

（2）getOutputStream()：获得一个输出流，用这个输出流将数据信息写入到网络"线路"上。

3．服务器端建立套接字 Socket 对象

编写 TCP 网络服务器程序时，要用 ServerSocket 类创建服务器 Socket，ServerSocket 类的构造方法如下：

```
ServerSocket(int port);
```

创建 ServerSocket 实例是不需要指定 IP 地址的，ServerSocket 总是处于监听本机端口的状态。

ServerSocket 类的主要方法如下：

```
Socket accept();
```

该方法用于在服务器端的指定端口监听客户机发起的连接请求，并与之连接，其返回值为 Socket 对象。

9.2 基于 TCP 的网络程序设计

基于 TCP 的网络程序采用的都是客户机/服务器系统模式。利用套接字 Socket 设计客户机/服务器系统程序进行数据通信与传输，大致有以下几个步骤：

（1）创建服务器端 ServerSocket，设置建立连接的端口号。
（2）创建客户端 Socket 对象，设置绑定的主机名称或 IP 地址，指定连接端口号。
（3）客户机 Socket 发起连接请求。
（4）建立连接。
（5）取得 InputStream 和 OutputStream。
（6）利用 InputStream 和 OutputStream 进行数据传输。
（7）关闭 Socket 和 ServerSocket。

客户机/服务器模式的连接请求与响应过程如图 9.5 所示。

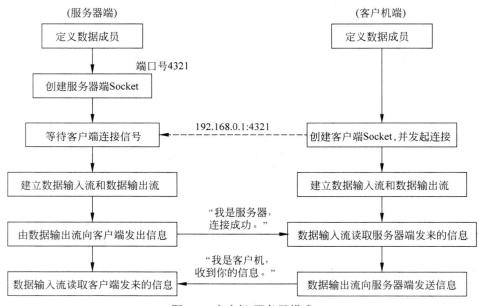

图 9.5 客户机/服务器模式

【例 9-2】远程数据通信示例，本例由 Android 客户端程序和 PC 机服务器程序两部分组成。

（1）Android 客户端程序代码如下：

```
1    package com.example.socketClient;
2    import java.io.DataInputStream;
3    import java.io.DataOutputStream;
4    import java.net.Socket;
5    import android.os.StrictMode;
6    import android.app.Activity;
```

视频演示

```
7   import android.os.Bundle;
8   import android.view.View;
9   import android.widget.Button;
10  import android.widget.TextView;

11  public class SocketClientActivity extends AppCompatActivity
12  {
13    private Socket socket=null;
14    private DataInputStream dis=null;
15    private DataOutputStream dos=null;
16    private TextView mTextView1;
17    private Button Button1;
18    String msg="";
19    @Override
20    public void onCreate(Bundle savedInstanceState)
21    {
22      super.onCreate(savedInstanceState);
23          setContentView(R.layout.main);
24      mTextView1=(TextView)findViewById(R.id.textView);
25      Button1=(Button) findViewById(R.id.Button);
26      Button1.setOnClickListener(new mClick());
27    //以下代码避免程序出现NetworkOnMainThreadException异常
28      StrictMode.setThreadPolicy(new  StrictMode
29              .ThreadPolicy
30              .Builder()
31              .detectDiskWrites()
32              .detectDiskReads()
33              .detectNetwork()
34              .penaltyLog()
35              .build() );
36    }
37   class mClick implements View.OnClickListener {
38    @Override
39    public void onClick(View v)
40    {
41       Client();
42    }
43    public void Client()
44    {
45  try {
46    socket=new Socket("192.168.0.1", 4321);
47      }catch (Exception ioe) {
48         System.out.print("socket  err ");
49      }
50  try{
```

设置线程策略

创建一个 socket 流连接到目标主机（请更换为目标主机的 IP）

```
51          //创建输入流对象dis读取数据，创建输出流对象dos发送数据
52           dis=new DataInputStream(socket.getInputStream());
53           dos=new DataOutputStream(socket.getOutputStream());
54           dos.writeUTF("给我数据啊。。。");
55           dos.flush();
56         } catch (IOException ioe) {
57            System.out.print("DataStream create err ");
58         }
59         ReadStr();          ◀── 读取服务器发来的数据
60        try{
61          Thread.sleep(500);
62            msg="手机客户端发来贺电! ";
63            WriteString(msg);  ◀── 发送数据
64         dis.close();
65         socket.close();
66          }catch (Exception ioe)
67       {  System.out.println("socket close() err ......."); }
68         ReadStr();   ◀── 读取服务器发来的第二条数据
69      }
70        //写数据到socket，即发送数据
71      public void WriteString(String str)
72      {
73     try {
74            dos.writeUTF(str);  ◀── 发送字符串str的数据
75            dos.flush();
76            ReadStr();
77            socket.close();
78         } catch (IOException e) {
79            System.out.print("WriteString() err");
80         }
81     }
82       //显示从socket返回的数据，即读取数据
83       public void ReadStr()
84       {
85        try {
86
87      if((msg=dis.readUTF()) != null)   ◀── 读取数据存放到字符串变量msg
88       {
89
90           mTextView1.append(msg);
91        }
92     } catch (IOException ioe) {
93     System.out.print("ReadStr() err ");
94        }
95    }
```

96   }

（2）服务器端程序代码如下：

```
1   import java.io.DataOutputStream;
2   import java.io.DataInputStream;
3   import java.io.IOException;
4   import java.net.ServerSocket;
5   import java.net.Socket;
6
7   public class server
8   {
9     private ServerSocket ss;
10    private Socket socket;
11    private DataInputStream dis;
12    private DataOutputStream dos;
13    public server()
14    {
15        new ServerThread().start();
16    }
17  class ServerThread extends Thread
18  {
19  public void run()
20  {
21    try {
22        ss=new ServerSocket(4321);           ◀── 实例化服务器端套接字对象
23        System.out.println("服务器启动了");
24        while(true)
25        {
26          socket=ss.accept();                ◀── 阻塞端口，等待客户机连接
27          System.out.println("有客户端连接到服务器");
28          dis=new DataInputStream(socket.getInputStream());
29          dos=new DataOutputStream(socket.getOutputStream());
30          dos.writeUTF("恭喜你，连接服务器成功！ \n");   ◀── 向手机发送一条数据
31          dos.flush();
32          System.out.println("服务器休眠20秒......");
33            Thread.sleep(500);
34          String msg="";
35        if((msg=dis.readUTF()) != null) {
36            System.out.println(msg);
37        }
38        dos.writeUTF("你发来的数据服务器收到了。^_^");   ◀── 发送第二条数据
39          dos.flush();
39        }
40    }
41        catch (Exception e) {System.out.println("读写错误");}
```

```
42        finally{
43            try {
44                in.close();
45                out.close();
46            } catch(IOException e) {e.printStackTrace();}
47        }
48   }
49 }
50   public static void main(String[] args) throws IOException
51   {
52       new server();
53   }
54 }
```

（3）在配置文件 AndroidManifest.xml 中加入允许访问网络的权限语句：

```
<uses-permission android:name="android.permission.INTERNET" />
```

【程序说明】
程序由客户机程序和服务器端程序两部分组成。
客户机程序：
（1）程序的第 28～35 行为设置线程策略，避免程序出现 NetworkOnMainThreadException 异常，这是 Android 与 Java 不同的地方。
（2）在第 46 行，创建一个可以连接到 Server 的套接字，其端口为 4321。运行程序时，当程序执行到该语句，立即向服务器发起连接。
（4）在第 59 行，调用 ReadStr()方法，通过数据输入流读取从服务器发送到"线路"里的信息。
（5）在第 63 行，调用 WriteString(String str)方法，通过数据输出流向由套接字建立的连接"线路"（向服务器端方向）发送信息。
服务器端程序：
（1）在第 17 行，创建多线程，可以用于多客户端的连接。
（2）在第 22 行，创建服务器端套接字，设定其端口号为 4321，该端口号与客户机套接字的端口号必须一致。注意，这里要使用 try-catch 结构处理异常事件。
（3）在第 26 行，服务器端套接字对象使用 accept()方法监听端口，等待接收客户机传来的连接信号。
（4）在第 28、29 行，建立套接字的数据输入流 dis 及数据输出流 dos。
（5）第 30 行，通过数据输出流向由套接字建立的连接"线路"（向客户机方向）发送"连接已经建立"的信息。
（6）在第 35 行，通过数据输入流读取客户机发送在"线路"里的信息。
（7）在第 37 行，显示接收到的信息。
将服务器端程序保存为 SServer.java，编译程序。首先运行服务器程序，然后再启动模拟器运行客户端程序。
程序运行结果如图 9.6 所示（先运行服务器端程序，再运行客户端程序）。

（a）客户端运行结果　　　　　　（b）服务器端运行结果

图 9.6　远程数据传输

9.3　基于 HTTP 的网络程序设计

HTTP（HyperText Transfer Protocol，超文本传输协议）是一个基于请求与响应模式的应用层协议，是基于 TCP 的连接方式，默认端口为 80。大多数的 Web 应用服务，都是构建在 HTTP 之上。

HTTP 的工作原理为，由 HTTP 的客户端发起请求，建立一个到服务器指定端口的 TCP 连接。HTTP 服务器则在那个端口监听客户端发送过来的请求。一旦收到请求，服务器（向客户端）发回一个响应信息，如"HTTP/1.1 200 OK"。

1. 请求方式

HTTP 协议有 GET 方式和 POST 方式两种请求方式。
- GET 方式：请求获取 Request-URI 所标识的资源，浏览器采用 GET 方式向服务器获取资源。
- POST 方式：在 Request-URI 所标识的资源后附加新的数据，常用于提交表单。

2. 请求信息

HTTP 请求报文由 3 个部分组成，分别是请求行、消息报头、请求正文。
HTTP 请求信息示例如下：

```
GET /hello.htm HTTP/1.1          ← 请求行，请求信息的标志
Accept: */*
Accept-Language: zh-cn
Accept-Encoding: gzip, deflate
If-Modified-Since: Wed, 17 Oct 2007 02:15:55 GMT
If-None-Match: W/"158-1192587355000"
User-Agent: Mozilla/4.0 (compatible; MSIE 6.0; Windows NT 5.1; SV1)
Host: 192.168.2.162:8080
Connection: Keep-Alive
```

HTTP 请求报文的头部参数见表 9-1。

表 9-1　HTTP 请求报文的头部参数

请求参数	参数说明
Accept	介质类型，*/* 表示任何类型
Accept-Charset	声明接收的字符集
Connection	保持连接的持续性，Keep-Alive 为保持连接的时间（秒）
Content-Encoding	压缩方式（gzip 或 deflate）
Content-Language	Web 服务器响应时使用的语言
Host	客户端指定访问 Web 服务器的域名/IP 地址和端口号
GET/HTTP/1.1	以 GET 方式请求，HTTP 的版本是 1.1
User-Agent	浏览器类型

3. HTTP 响应信息

HTTP 响应信息的报文示例如下：

```
HTTP/1.1 200 OK          ← 响应信息的标志
Last-Modified: Wed, 17 Oct 2007 03:01:41 GMT
Content-Type: text/html
Content-Length: 158
Date: Wed, 17 Jul 2012 03:01:59 GMT
Server: Apache-Coyote/1.1
```

HTTP 响应报文的头部参数见表 9-2。

表 9-2　HTTP 协议响应报文的头部参数

响应参数	参数说明
Date	当前响应的 GMT 时间
Connection	保持连接的持续性
Content-Length	响应文档的长度，以字节方式存储的十进制数表示
Content-Type	Web 服务器响应文档的 MIME 类型
Cache-Control	缓存的控制权限
Content-Encoding	文档编码的压缩方式（gzip 或 deflate）
Expires	文档已过期时间
HTTP/1.1 200 OK	HTTP 的版本是 1.1，返回 200 表示成功
Server	Web 服务器系统及版本等信息

【例 9-3】显示 HTTP 报文头部信息。

在该项目的界面设计中，设置一个文本编辑框和一个按钮。在按钮的事件中，通过套接字 Socket 建立的输出流向 www.baidu.com 网站发出访问请求信息，通过套接字 Socket 建立的输入流接收网站发来的响应报文，最后将接收到的报文头部信息显示到文本编辑框中。

视频演示

其代码如下：

```
1    package com.example.ex09_03;
2    import java.io.BufferedReader;
3    import java.io.InputStream;
4    import java.io.InputStreamReader;
5    import java.io.OutputStream;
```

```
6   import java.net.Socket;
7   import android.os.Bundle;
8   import android.view.View;
9   import android.view.View.OnClickListener;
10  import android.widget.Button;
11  import android.widget.TextView;
12  import android.app.Activity;
13
14  public class MainActivity extends Activity
15  {
16    TextView text = null;
17    Button httpBtn;
18    @Override
19    public void onCreate(Bundle savedInstanceState)
20    {
21        super.onCreate(savedInstanceState);
22        setContentView(R.layout.activity_main);
23        text = (TextView) this.findViewById(R.id.textView1);
24        httpBtn=(Button)findViewById(R.id.button1);
25        httpBtn.setOnClickListener(new mClick());
26    }
27    class mClick implements OnClickListener
28    {
29      @Override
30      public void onClick(View arg0)
31      {
32         String host="www.baidu.com";
33         String url="/index.html";
34         String method="GET";
35        StringBuffer sb1, sb2;
36        String str;
37        OutputStream outStream;
38        InputStream inStream;
39        InputStreamReader inReader;
40        BufferedReader buff;
41        try {
42          Socket socket=new Socket(host, 80);     ← 实例化 Socket 套接字对象
43          outStream=socket.getOutputStream();     ← 建立输出流对象
44          sb1=new StringBuffer();
45          /**
46           * 请求信息第 1 行: 方式, 请求的内容, HTTP 协议的版本
47           * 用 GET 方式, 请求的内容是 url, HTTP 协议的版本为 1.1 版 "HTTP/1.1"
48           **/
49          sb1.append(method + " " + url + " HTTP/1.1\r\n");
50          /* 请求信息的第 2 行: 主机名, 格式为 "Host:主机"    */
51          sb1.append("Host:" + host + "\r\n");
52          /* 请求信息的第 3 行: 接收的数据类型            */
53          sb1.append("Accept: :*/* \r\n");
```

```
54        /*请求信息的第 4 行: 连接设置设定为一直保持连接*/
55        sb1.append("Connection: Keep-Alive\r\n");
56        /*请求信息的第 5 行: 注意最后一定要有\r\n回车换行*/
57        sb1.append("\r\n");
58        outStream.write(sb1.toString().getBytes());     ← 发送请求报文
59        outStream.flush();
60
61        inStream=socket.getInputStream();               ← 建立输入流对象
62        inReader=new InputStreamReader(inStream);
63        buff=new BufferedReader(inReader);
64        sb2=new StringBuffer();
65        while((str=buff.readLine()) != null)            ← 从网络"线路"上
66        {                                                  读取数据
67            sb2.append(str + "\n");
68        }
69        buff.close();
70        inReader.close();
71        outStream.close();
72        inStream.close();
73        text.setText(sb2.toString());                   ← 将读取的内容放到文本框中显示
74        } catch (Exception e) {
75            System.out.println("套接字连接错误," + e);
76        }
77    }
78  }
79 }
```

在配置文件 AndroidManifest.xml 中加入允许访问网络的权限语句:

`<uses-permission android:name= "android.permission.INTERNET" />`

程序的运行结果如图 9.7 所示。

图 9.7　显示 HTTP 报文头部信息

9.4 Web 视图

9.4.1 浏览器引擎 WebKit

WebKit 是一个开源的浏览器引擎。WebKit 内核具有非常好的网页解析机制，很多应用系统都使用 WebKit 作为浏览器的内核。例如，Google 的 Android 系统、Apple 的 iOS 系统、Nokia 的 Series 60 Browser 系统所使用的 Browser 内核引擎，都是基于 WebKit 的。WebKit 所包含的 WebCore 排版引擎和 JSCore 引擎来自于 KDE 的 KHTML 和 KJS，它们拥有清晰的源码结构和极快的渲染速度。

Android 对 Webkit 做了进一步封装，并提供了丰富的 API。Android 平台的 WebKit 模块由 Java 层和 WebKit 库两个部分组成，Java 层负责与 Android 应用程序进行通信，而 WebKit 类库负责实际的网页排版处理。WebKit 包中的几个重要类见表 9-3。

表 9-3 WebKit 包中的几个重要类

类 名	说 明
WebSettings	用于设置 WebView 的特征、属性等
WebView	显示 Web 页面的视图对象，用于网页数据的载入、显示等操作
WebViewClient	在 Web 视图中帮助处理各种通知、请求事件
WebChromeClient	Google 浏览器 Chrome 的基类，辅助 WebView 处理 JavaScript 对话框、网站的标题、网站的图标、加载进度条等

9.4.2 Web 视图对象

1. WebView 类

在 WebKit 的 API 包中，最重要、最常用的类是 Android.WebKit.WebView。WebView 类是 WebKit 模块 Java 层的视图类，所有需要使用 Web 浏览功能的 Android 应用程序都要创建该视图对象，用于显示和处理请求的网络资源。目前，WebKit 模块支持 HTTP、HTTPS、FTP 及 JavaScript 请求。WebView 作为应用程序的 UI 接口，为用户提供了一系列的网页浏览、用户交互接口，客户程序通过这些接口访问 WebKit 核心代码。

WebView 类的常用方法见表 9-4。

表 9-4 WebView 类的常用方法

方 法	说 明
WebView(Context context)	构造方法
loadUrl(String url)	加载 URL 网站页面
loadData(String data, String mimeType, String encod)	显示 HTML 格式的 Web 视图
reload()	重新加载网页
getSettings()	获取 WebSettings 对象
goBack()	返回上一个页面

方　法	说　明
goForward()	向前一个页面
clearHistory()	清除历史记录
addJavascriptInterface (Object obj, String interfaceName)	将对象绑定到 JavaScript，允许从网页控制 Android 程序，从网页调用该对象的方法

2. 使用 WebView 的说明

（1）设置 WebView 的基本信息：

- 如果访问的页面中有 JavaScript，则 WebView 必须设置支持 JavaScript。

```
webview.getSettings().setJavaScriptEnabled(true);
```

- 触摸焦点起作用：

```
requestFocus();
```

- 取消滚动条：

```
this.setScrollBarStyle(SCROLLBARS_OUTSIDE_OVERLAY);
```

（2）设置 WebView 要显示的网页：

- 互连网用 webView.loadUrl("http://www.google.com");
- 本地文件用 webView.loadUrl("file:///android_asset/XX.html");

注意，本地文件要存放在项目的 assets 目录中。

（3）用 WebView 单击链接看了很多页面以后，如果不做任何处理，按 Backspace 键，浏览器会调用 finish()结束自身的运行；如果希望浏览的网页回退而不是退出浏览器，需要在当前 Activity 中覆盖 Activity 类的 onKeyDown(int keyCoder,KeyEvent event)方法处理该 Back 事件。

```
public boolean onKeyDown(int keyCoder,KeyEvent event)
{
 if(webView.canGoBack() && keyCoder==KeyEvent.KEYCODE_BACK)
  {
    webview.goBack();      ← goBack()表示返回 WebView 的上一个页面
    return true;
  }
  return false;
}
```

【例 9-4】 应用 WebView 对象浏览网页。

（1）设计界面布局文件 activity_main.xml。在界面布局中，设置了一个文本编辑框，用于输入网址；设置了一个按钮，用于打开网页；还设置了一个网页视图组件 WebView，用于显示网页。

其代码如下：

```
1   <?xml version="1.0" encoding="utf-8"?>
2   <LinearLayout xmlns:android="http://schemas.android.com/apk/res/android"
3       android:layout_width="fill_parent"
```

```
4       android:layout_height="fill_parent"
5       android:layout_gravity="center_horizontal"
6       android:orientation="vertical" >
7    <LinearLayout
8        android:id="@+id/LinearLayout2"
9        android:layout_width="fill_parent"
10       android:layout_height="wrap_content" >
11       <EditText
12           android:id="@+id/editText1"
13           android:layout_width="207dp"
14           android:layout_height="wrap_content"/>
15       <Button
16           android:id="@+id/button1"
17           android:layout_width="wrap_content"
18           android:layout_height="wrap_content"
19           android:layout_weight="1"
20           android:text="打开网页" />
21   </LinearLayout>
22   <WebView
23       android:id="@+id/webView1"
24       android:layout_width="fill_parent"
25       android:layout_height="fill_parent" />
26 </LinearLayout>
```

视频演示

（2）控制文件 MainActivity.java 的代码如下：

```
1  package com.example.ex09_04;
2  import android.os.Bundle;
3  import android.app.Activity;
4  import android.view.View;
5  import android.view.View.OnClickListener;
6  import android.webkit.WebView;
7  import android.widget.Button;
8  import android.widget.EditText;
9
10 public class MainActivity extends Activity
11 {
12   WebView webView;
13   Button openWebBtn;
14   EditText edit;
15   @Override
16   public void onCreate(Bundle savedInstanceState)
17   {
18       super.onCreate(savedInstanceState);
19       setContentView(R.layout.activity_main);
20       openWebBtn=(Button)findViewById(R.id.button1);
21       edit=(EditText)findViewById(R.id.editText1);
22       openWebBtn.setOnClickListener(new mClick());
```

```
23     }
24     class mClick implements OnClickListener
25     {
26       public void onClick(View arg0)
27       {
28           String url=edit.getText().toString();
29           webView=(WebView)findViewById(R.id.webView1);
30           webView.loadUrl("http://" + url);
31       }
32     }
33   }
```

在配置文件 AndroidManifest.xml 中加入允许访问网络的权限语句：

```
<uses-permission android:name="android.permission.INTERNET" />
```

程序的运行结果如图 9.8 所示。

图 9.8　用 WebView 显示网页

9.4.3　调用 JavaScript

1. WebSteeings 类

WebView 对象刚创建时，使用的是系统默认设置，当需要对 WebView 对象的属性等进行自定义设置时，需要用到 WebSteeings 类。WebSteeings 类的常用方法见表 9-5。

表 9-5　WebSteeings 类的常用方法

方　　法	说　　明
setAllowFileAccess(boolean flag)	设置是否允许访问文件数据
setJavaScriptEnabled(boolean flag)	设置是否支持 JavaScript 脚本
setBuiltInZoomControls(boolean flag)	设置是否支持缩放
setBlockNetworkImage (boolean flag)	设置是否禁止显示图片，true 为禁止显示
setDefaultFontSize (int size)	设置默认字体大小，在 1～72 取值
setTextZoom (int textZoom)	设置页面文字缩放的百分比，默认为 100

2. WebViewClient 类

WebViewClient 类用于对 WebView 对象中的各种事件的处理，通过重写这些提供的事件方法，可以对 WebView 对象在页面载入、资源载入、页面访问错误等情况发生时进行各种操作。WebViewClient 类的常用方法见表 9-6。

表 9-6 WebViewClient 类的常用方法

方法	说明
onLoadResource(WebView view, String url)	通知 WebView 加载 url 指定的资源时触发
onPageStarted(WebView view, String url, Bitmap favicon)	页面开始加载时触发
onPageFinished(WebView view, String url)	页面加载完毕时触发

3. WebChromeClient 类

WebChromeClient 是辅助 WebView 处理 JavaScript 对话框、网站的标题、网站的图标、加载进度条等操作的类，其常用方法见表 9-7。

表 9-7 WebChromeClient 类的常用方法

方法	说明
onJsAlert (WebView view, String url, String message, JsResult result)	处理 JavaScript 的 alert 对话框
onJsPrompt (WebView view, String url, String message, String defaultValue, JsPromptResult result)	处理 JavaScript 的 Prompt 提示对话框
onCloseWindow (WebView window)	关闭 WebView

【例 9-5】在 Android 程序中调用 HTML 代码示例。

（1）界面布局文件的代码如下：

```
1   <LinearLayout xmlns:android="http://schemas.android.com/
    apk/res/android"
2       xmlns:tools="http://schemas.android.com/tools"
3       android:id="@+id/LinearLayout1"
4       android:layout_width="fill_parent"
5       android:layout_height="fill_parent"
6       android:orientation="vertical" >
7       <TextView
8           android:id="@+id/textView1"
9           android:layout_width="wrap_content"
10          android:layout_height="wrap_content"
11          android:layout_marginTop="10dp"
12          android:padding="@dimen/padding_medium"
13          android:text="@string/hello_world"
14          tools:context=".MainActivity" />
15      <Button
16          android:id="@+id/button1"
17          android:layout_width="wrap_content"
18          android:layout_height="wrap_content"
19          android:layout_marginTop="10dp"
```

视频演示

```
20          android:text="Button" />
21      <WebView
22          android:id="@+id/WebView1"
23          android:layout_width="fill_parent"
24          android:layout_height="fill_parent" />
25  </LinearLayout>
```

（2）控制文件的代码如下：

```
1   package com.example.web1;
2   import android.os.Bundle;
3   import android.app.Activity;
4   import android.view.View;
5   import android.view.View.OnClickListener;
6   import android.webkit.WebChromeClient;
7   import android.webkit.WebSettings;
8   import android.webkit.WebView;
9   import android.widget.Button;
10
11  public class MainActivity extends Activity
12  {
13      Button webBtn;
14      WebView web;
15      @Override
16      public void onCreate(Bundle savedInstanceState)
17      {
18          super.onCreate(savedInstanceState);
19          setContentView(R.layout.activity_main);
20          web=(WebView)findViewById(R.id.WebView1);
21          webBtn=(Button)findViewById(R.id.button1);
22          webBtn.setOnClickListener(new mClick());
23      }
24      class mClick implements OnClickListener
25      {
26          public void onClick(View arg0)
27          {
28              String summary =
29                "<html>
30                    <body>
31                        You scored <b> 96 </b> points.     ← HTML 代码
32                    </body>
33                </html>";
34              WebSettings setting=web.getSettings();
35              setting.setJavaScriptEnabled(true);
36              web.setWebChromeClient(new WebChromeClient());
37              web.loadData(summary, "text/html", "utf-8");
```

```
38        }
39    }
40 }
```

在配置文件 AndroidManifest.xml 中加入允许访问网络的权限语句：

```
<uses-permission android:name="android.permission. INTERNET" />
```

程序的运行结果如图 9.9 所示。

【**例 9-6**】调用 JavaScript 程序示例。

（1）在项目的 assets 下，新建一个 JavaScript 程序 test.html。assets 存放应用程序使用的外部资源文件，res 存放应用程序的资源文件。JavaScript 程序 test.html 的代码如下：

图 9.9　运行 HTML 代码

```
1  <HTML>
2  <head>
3      <title> 一个简单的 JavaScript 示例  </title>
4  </head>
5  <body>
6  <script language="javascript" type="text/javascript" >
7      function addAll(a, b, c)
8      {
9          return a+b+c;
10     }
11     var total=addAll(30, 40, 50);
12     var str="Ran 5 hours,<br>  finally finished the  ";
13     document.write("<html><B> " + str + total + "  km!</B></html>");
14 </script>
15 </body>
16 </HTML>
```

视频演示

（2）界面布局文件的代码如下：

```
1  <RelativeLayout xmlns:android="http://schemas.android.com/apk/res/android"
2      xmlns:tools="http://schemas.android.com/tools"
3      android:layout_width="fill_parent"
4      android:layout_height="fill_parent" >
5      <WebView
6          android:id="@+id/webView1"
7          android:layout_width="fill_parent"
8          android:layout_height="fill_parent"  />
9  </RelativeLayout>
```

（3）控制文件的代码如下：

```
1  package com.example.ex09_06;
2  import android.os.Bundle;
3  import android.webkit.WebSettings;
4  import android.webkit.WebView;
5  import android.app.Activity;
6
```

```
7   public class MainActivity extends Activity
8   {
9       private WebView webView=null;
10      private WebSettings webSettings=null;
11      @Override
12      public void onCreate(Bundle savedInstanceState)
13      {
14          super.onCreate(savedInstanceState);
15          setContentView(R.layout.activity_main);
16          webView=(WebView) this.findViewById(R.id.webView1);
17          webSettings=webView.getSettings();
18          webSettings.setAllowFileAccess(true);          ← 设置允许访问文件数据
19          webSettings.setJavaScriptEnabled(true);        ← 设置支持 JavaScript 脚本
20          webSettings.setBuiltInZoomControls(true);      ← 设置支持缩放
21          webView.loadUrl("file:///android_asset/test.html");
22      }
23  }
```

程序的运行结果如图 9.10 所示。

【例 9-7】用 Android 程序操纵 JavaScript 对话框。

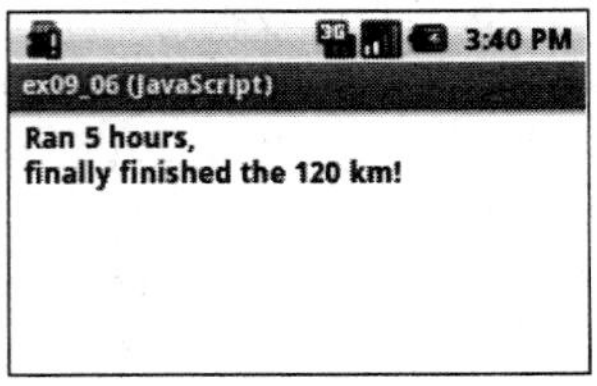

（1）在项目的 assets 下，新建一个 JavaScript 对话框程序 test1.html，其代码如下：

图 9.10　调用 JavaScript 示例

```
1   <html>
2   <head>
3       <title>JavaScript 与 Android 交互</title>
4   </head>
5   <script type="text/javascript">
6    function show_alert()
7    {
8     var a=document.getElementById("text").value;
9     alert("Hello " + a );
10   }
11  </script>
12  <body>
13  <form action="">
14      <input type="text" id="text" value=""/>
15      <input type="button" id="button"
16             onclick="window.test.android_show()"    ← 调用 Android 程序标记为 test 的实例对象的函数
17             value="call Android"/>
18  </form>
19  </body>
20  </html>
```

视频演示

JavaScript 对话框

（2）界面布局文件的代码同例 9-6。

（3）控制文件的代码如下：

```java
1   package com.example.ex09_07;
2   import android.os.Bundle;
3   import android.os.Handler;
4   import android.webkit.JsResult;
5   import android.webkit.WebChromeClient;
6   import android.webkit.WebSettings;
7   import android.webkit.WebView;
8   import android.widget.Toast;
9   import android.app.Activity;
10  
11  public class MainActivity extends Activity
12  {
13    WebView webView;
14    Handler handler=new Handler();
15    MWebChromeClient mWebChromeClient;
16    @Override
17    public void onCreate(Bundle savedInstanceState)
18    {
19      super.onCreate(savedInstanceState);
20      setContentView(R.layout.activity_main);
21      webView=(WebView) findViewById(R.id.webView1);
22      WebSettings webSettings=webView.getSettings();
23      webSettings.setAllowFileAccess(true);          ← 设置允许访问文件数据
24      webSettings.setJavaScriptEnabled(true);        ← 设置支持 JavaScript 脚本
25      webSettings.setBuiltInZoomControls(true);      ← 设置支持缩放
26      webSettings.setDefaultFontSize(24);
27      MObject mObject=new MObject();
28      webView.addJavascriptInterface(mObject, "test");
29      mWebChromeClient=new MWebChromeClient();
30      webView.setWebChromeClient(mWebChromeClient);
31      webView.loadUrl("file:///android_asset/test1.html");
32    }
33    class MObject extends Object
34    {
35      public void android_show()
36      {
37        handler.post(new Runnable()
38        {
39          public void run()
40          {
41            System.out.println("提示: 调用了多线程的 run()方法!!");
42            webView.loadUrl("javascript: show_alert()");  ← 调用 JavaScript 函数
43          }
44        };
45    }
```

```
46      }
47      class MWebChromeClient extends WebChromeClient
48      {
49          @Override
50          public boolean onJsAlert(WebView view,          ← 处理 JavaScript 的 alert 对话框
51                  String url, String message, JsResult result)
52          {
53              Toast.makeText(getApplicationContext(), message,
54                      Toast.LENGTH_LONG).show();
55              return true;
56          }
57      }
58  }
```

在配置文件 AndroidManifest.xml 中加入允许访问网络的权限语句：

`<uses-permission android:name="android.permission.INTERNET" />`

程序的运行结果如图 9.11 所示。

图 9.11 操纵 JavaScript 对话框

【例 9-8】与 JavaScript 程序交互，实现单击网页上的图片后，切换到另一图片的功能。

（1）复制两个图片文件 a.jpg 和 b.jpg 到项目的 assets 下，然后在该目录下新建一个 JavaScript 网页文件 test2.html，实现单击图片后切换到另一图片的功能。其代码如下：

```
1   <html>
2     <script language="javascript">
3       function loadPhoto()              ← 定义函数，Android 程序调用该函数
4       {
5         document.getElementById("droid").src="b.jpg";   ← 在网页中加载图片
6       }
7     </script>
8     <body>
9       <a onClick="window.aaa.clickOnAndroid()">    ← 调用 Android 程序标记为 aaa 的实例对象的函数
10        <img id="droid" src="a.jpg"/><br>
11        Click me!
```

```
12        </a>
13     </body>
14  </html>
```

（2）界面布局文件的代码同例 9-6。

（3）控制文件的代码如下：

```
1   package com.example.ex09_08;
2   import android.os.Bundle;
3   import android.os.Handler;
4   import android.app.Activity;
5   import android.webkit.WebSettings;
6   import android.webkit.WebView;
7
8   public class MainActivity extends Activity
9   {
10    private WebView mWebView;
11    private Handler mHandler = new Handler();
12    @Override
13    public void onCreate(Bundle savedInstanceState)
14    {
15      super.onCreate(savedInstanceState);
16      setContentView(R.layout.activity_main);
17      mWebView=(WebView) findViewById(R.id.webView1);
18      WebSettings webSettings=mWebView.getSettings();
19      webSettings.setJavaScriptEnabled(true);      ← 设置支持 JavaScript 脚本
20      mWebView.addJavascriptInterface(new mObject(), "aaa");  ← 将对象绑定到 JavaScript，允许
21      mWebView.loadUrl("file:///android_asset/test2.html");      从网页调用该对象的方法，
22    }                                                            aaa 为对象标记
23    class mObject extends Object
24    {
25      public void clickOnAndroid()
26      {
27          mHandler.post(new mRunnable());    ← 网页文件调用该函数，实现图片的加载
28      }
29    }
30    class mRunnable implements Runnable
31    {
32      public void run()
33      {
34          mWebView.loadUrl("javascript: loadPhoto()");  ← 调用网页文件的函数，加载另一幅图片
35      }
36   }
37 }
```

程序的运行结果如图 9.12 所示。

图 9.12　单击网页上的图片，切换到另一幅图片

9.5　无线网络通信技术 WiFi

WiFi（Wireless Fidelity，WiFi，读作"wai－fai"）是一种无线网络通信技术，它的传输速度可以达到 11Mb/s。由于 WiFi 频段在世界范围内是无须任何电信运营执照的免费频段，因此，WLAN 无线设备提供了一个世界范围内可以使用的、费用极低且数据带宽极高的无线空中接口。用户可以在 WiFi 覆盖区域内快速浏览网页，随时随地接听拨打电话。对于其他一些基于 WLAN 的宽带数据应用，如流媒体、网络游戏等功能，更是值得用户期待。有了 WiFi 功能，打长途电话（包括国际长途）、浏览网页、收发电子邮件、下载音乐、传递数码照片等，再无须担心速度慢和花费高的问题。

1. wifi 包

Android 系统提供的实现 WiFi 无线网络通信技术的 API 均在 android.net.wifi 包中，其主要类见表 9-8。

表 9-8　WiFi 无线网络通信技术常用的 API

类　　名	说　　明
ScanResult	主要用来描述已经检测出的接入点，包括接入点的地址、接入点的名称、身份认证、频率、信号强度等信息
WifiConfiguration	WiFi 网络的配置，包括安全设置等
WifiInfo	描述 WiFi 的连接信息
WifiManager	管理 WiFi 连接

2. WifiInfo 类

WifiInfo 类的功能主要是描述 WiFi 的无线网络连接信息，其信息内容包括接入点、网络连接状态、IP 地址、连接速度、MAC 地址、无线网络的 ID、信号强度等。WifiInfo 类的主要方法见表 9-9。

表 9-9　WifiInfo 类的主要方法

方　　法	说　　明
getIpAddress()	获取 IP 地址
getLinkSpeed()	获取连接的速度
getMacAddress()	获取 Mac 地址
getRssi()	获取 802.11n 网络信号
getSSID()	获取无线网络的 SSID（SSID 为区分不同网络的标识）
getBSSID()	获取网络的 BSSID（BSSID 为无线网络站点的 MAC 地址）

3. WifiManager 类

WifiManager 是管理 WiFi 连接的类，其主要方法见表 9-10。

表 9-10　WifiManager 类的主要方法

方　　法	说　　明
addNetwork(WifiConfiguration config)	通过获取到的网络连接信息来添加网络
calculateSignalLevel(int rssi, int numLevels)	计算信号的等级
createMulticastLock(int lockType, String tag)	创建 WiFi 锁，锁定当前的 WiFi 连接
disconnect()	断开连接
enableNetwork(int netId, boolean disableOthers)	允许与一个网络进行连接
getConfiguredNetworks()	获取网络连接的状态
getConnectionInfo()	获取当前连接的信息
getScanResults()	获取扫描结果
getWifiState()	获取 WiFi 的功能状态
pingSupplicant()	检查对请求的响应状况，判断是否连通
reconnect()	重新连接当前断开的接入点
startScan()	开始扫描

对于 WifiManager 来说，经常使用以下操作：

1）得到当前网卡状态

```
int wifiState = wifiManger.getWifiState();
```

这个函数返回的是一个整型数值，不同的返回值代表不同的状态，每一种状态都对应着一个常量，这些常量存放在 WifiManager 类中，其具体含义见表 9-11。

表 9-11　网卡状态常量

常　　量	返　回　值	说　　明
WIFI_STATE_DISABLED	1 (0x00000001)	WiFi 网卡不可用
WIFI_STATE_DISABLING	0 (0x00000000)	WiFi 正在关闭
WIFI_STATE_ENABLED	3 (0x00000003)	WiFi 网卡可用
WIFI_STATE_ENABLING	2 (0x00000002)	WiFi 网卡正在打开
WIFI_STATE_UNKNOWN	4 (0x00000004)	未知网卡状态

2）改变当前网卡状态

通过 WifiManager 对象可以改变 WiFi 网卡的状态。

- 打开 WiFi 网卡：

```
wifiManager.setWifiEnabled(true);
```

- 关闭 WiFi 网卡：

```
wifiManager.setWifiEnabled(false);
```

需要说明的是，WiFi 网卡的打开和关闭并不是瞬间的过程，需要一段时间。也就是说，如果当前手机的网卡处于可用状态，关闭网卡之后，并不马上进入关闭状态，而是处于正在关闭状态，等关闭的动作完成以后才会真正进入关闭状态。

4. 操作 WiFi 所需要的权限

在应用程序中使用 WiFi 网络通信，需要在配置文件 AndroidManifest.xml 中加入允许访问 WiFi 的权限语句，见表 9-12。

表 9-12 操作 WiFi 所需要的权限

需要的权限	说　　明
CHANGE_NETWORK_STATE	修改网络状态的权限
CHANGE_WIFI_STATE	修改 WiFi 状态的权限
ACCESS_NETWORK_STATE	访问网络的权限
ACCESS_WIFI_STATE	访问 WiFi 的权限

【例 9-9】 测试 WiFi 网络。

（1）在项目中新建一个连接，并测试 WiFi 网络的程序文件 WifiTest.java，其代码如下：

```
1   package com.example.ex09_09;
2   import java.util.List;
3   import android.content.Context;
4   import android.net.wifi.ScanResult;
5   import android.net.wifi.WifiConfiguration;
6   import android.net.wifi.WifiInfo;
7   import android.net.wifi.WifiManager;
8   import android.net.wifi.WifiManager.WifiLock;
9
10  public class WifiTest
11  {
12      private WifiManager mWifiManager;          ← 定义 WifiManager 对象
13      private WifiInfo mWifiInfo;                ← 定义 WifiInfo 对象
14      private List<ScanResult> mWifiList;        ← 扫描出的网络连接列表
15      private List<WifiConfiguration> mWifiConfiguration;   ← 网络连接列表
16      WifiLock mWifiLock;                        ← 定义一个 WifiLock
17      public WifiTest(Context context)           ← 构造方法
18      {
19          mWifiManager=(WifiManager) context
20              .getSystemService(Context.WIFI_SERVICE);   ← 取得 WifiManager 对象
21          mWifiInfo=mWifiManager.getConnectionInfo();    ← 取得 WifiInfo 对象
```

```
22    }
23    /*打开 WiFi*/
24    public void openWifi()
25    {
26    if(!mWifiManager.isWifiEnabled())
27        {
28        mWifiManager.setWifiEnabled(true);
29    }
30    }
31    /*关闭 WiFi*/
32    public void closeWifi()
33    {
34      if(mWifiManager.isWifiEnabled())
35        {
36          mWifiManager.setWifiEnabled(false);
37        }
38    }
39    /*检查当前 WiFi 状态*/
40    public int checkState()
41    {
42        return mWifiManager.getWifiState();
43    }
44    /*检查网络*/
45    public void startScan()
46    {
47      mWifiManager.startScan();
48      mWifiList = mWifiManager.getScanResults();        ◀── 得到扫描结果
49      mWifiConfiguration =
50          mWifiManager.getConfiguredNetworks();         ◀── 得到配置好的网络连接
51    }
52    /*得到网络列表*/
53    public List<ScanResult> getWifiList()
54    {
55      return mWifiList;
56    }
57 }
```

（2）主控程序 MainActivity.java 的代码如下：

```
1    package com.example.ex09_09;
2    import java.util.List;
3    import android.app.Activity;
4    import android.net.wifi.ScanResult;
5    import android.os.Bundle;
6    import android.view.View;
7    import android.view.View.OnClickListener;
```

```java
8   import android.widget.Button;
9   import android.widget.ScrollView;
10  import android.widget.TextView;
11  import android.widget.Toast;
12
13  public class MainActivity extends Activity implements OnClickListener
14  {
15      private ScrollView sView;            ← 右侧滚动条按钮
16      private TextView allNetWork;
17      private Button scan;
18      private Button start;
19      private Button stop;
20      private Button check;
21      private WifiTest mWifiAdmin;
22      private List<ScanResult> list;       ← 扫描出的网络连接列表
23      private ScanResult mScanResult;
24      private StringBuffer mStringBuffer = new StringBuffer();
25      @Override
26      public void onCreate(Bundle savedInstanceState)
27      {
28          super.onCreate(savedInstanceState);
29          setContentView(R.layout.activity_main);
30          mWifiAdmin = new WifiTest(this);
31          init();
32      }
33      /*按钮的初始化*/
34      public void init()
35      {
36          sView=(ScrollView) findViewById(R.id.mScrollView);
37          allNetWork=(TextView)findViewById(R.id.allNetWork);
38          scan=(Button) findViewById(R.id.scan);
39          start=(Button) findViewById(R.id.start);
40          stop=(Button) findViewById(R.id.stop);
41          check=(Button) findViewById(R.id.check);
42          scan.setOnClickListener(this);
43          start.setOnClickListener(this);
44          stop.setOnClickListener(this);
45          check.setOnClickListener(this);
46      }
47      /*打开*/
48      public void start()
49      {
50          mWifiAdmin.openWifi();
51          Toast.makeText(this, "当前Wifi网卡状态为" + mWifiAdmin.checkState(),
52                  Toast.LENGTH_SHORT).show();
```

```
53        }
54     /*关闭*/
55    public void stop()
56    {
57        mWifiAdmin.closeWifi();
58        Toast.makeText(this, "当前Wifi网卡状态为" + mWifiAdmin.checkState(),
59             Toast.LENGTH_SHORT).show();
60    }
61     /*检查状态*/
62    public void check()
63    {
64        Toast.makeText(this, "当前Wifi网卡状态为" + mWifiAdmin.checkState(),
65             Toast.LENGTH_SHORT).show();
66    }
67     /*扫描网络*/
68    public void getAllNetWorkList()
69    {
70        //每次扫描之前清空上一次的扫描结果
71        if (mStringBuffer != null)
72        {
73            mStringBuffer = new StringBuffer();
74        }
75        //开始扫描网络
76        mWifiAdmin.startScan();
77        list=mWifiAdmin.getWifiList();
78        if (list!=null)
79        {
80            for (int i=0; i < list.size(); i++)
81            {
82                mScanResult=list.get(i);
83                mStringBuffer=mStringBuffer        ← 得到网络的SSID
84                    .append(mScanResult.SSID).append("    ")
85                    .append(mScanResult.BSSID).append("    ")
86                    .append(mScanResult.capabilities).append("    ")
87                    .append(mScanResult.frequency).append("    ")
88                    .append(mScanResult.level).append("    ")
89                    .append("\n\n");
90            }
91            allNetWork.setText("扫描到的所有Wifi网络: \n"
92                 + mStringBuffer.toString());
93        }
94    }
95
96    public void onClick(View v)
97    {
```

```
 98        switch (v.getId())
 99        {
100        case R.id.scan:
101            getAllNetWorkList();
102            break;
103        case R.id.start:
104            start();
105            break;
106        case R.id.stop:
107            stop();
108            break;
109        case R.id.check:
110            check();
111            break;
112        default:
113            break;
114        }
115    }
116 }
```

在配置文件 AndroidManifest.xml 中加入允许使用 wifi 访问网络所需的权限语句：

图 9.13　在真实手机上运行程序的结果

```
<uses-permission android:name="android.permission.CHANGE_NETWORK_STATE">
</uses-permission>
 <uses-permission android:name="android.permission.CHANGE_WIFI_STATE">
</uses-permission>
 <uses-permission android:name="android.permission.ACCESS_NETWORK_STATE">
</uses-permission>
 <uses-permission android:name="android.permission.ACCESS_WIFI_STATE">
</uses-permission>
```

WiFi 不能在模拟器上测试，需要在实体设备上才能运行。图 9.13 为在真实手机上运行程序的截图。

习　题　9

1. 编写一个可以发送和接收文本内容的聊天室程序。
2. 编写一个 JavaScript 网站，通过 Android 访问并操控该网站。
3. 编写一个 WiFi 传送图片的网络应用程序。

第 10 章　地图服务及传感器检测技术

10.1　Google 地图

Google 地图（Google Maps）是 Google 公司提供的电子地图服务，包括普通电子地图、交通地图和详细的卫星地图等。Android 作为 Google 公司旗下的产品，当然具备了 Google 地图的所有优秀功能。下面介绍 Android 系统中 Google 地图与应用项目整合的方法。

10.1.1　Google Maps 包

Google Maps 包不是 Android 系统 SDK 的标准库，标准的 Android 系统 SDK 中并未包含 Google Maps 包。因此，在创建基于 Google 地图的应用程序时，需要将相应的 Google 地图包加入到应用项目中。Google Maps 包位于 Android 系统 SDK 安装目录的 "add-ons\addon-google_apis-google_inc_-（API 版本号）\libs" 下，其中，API 版本号为 API 版本所代表的数字，例如，Android 4.03 的地图包就位于 SDK 安装目录下的 "add-ons\addon-google_apis-google_inc_-15\libs" 中。

Google Maps 包的包名为 com.google.android.map，其中包含了一系列用于在 Google 地图上显示、控制和层叠信息的功能类，表 10-1 是该包中几个重要的类。

表 10-1　Google Maps 包中的重要类

类　名	说　明
MapActivity	用于显示 Google 地图的 Activity 类，它需要连接底层网络
MapView	用于显示地图的 View 组件，必须和 MapView 配合使用
MapController	用于控制地图的移动、缩放等
OverLay	覆盖在 MapView 的图层，可显示位于地图之上的可绘制的对象
GeoPoint	包含经纬度位置的对象

下面对 Google Maps 包中的几个重要类进行简要说明。

（1）MapActivity：抽象类，继承于 Activity 类，用于处理显示 Google 地图所需要的服务。任何显示 Google 地图的 Activity 视图都必须继承它，并在 onCreate()方法中创建 MapView 对象的实例。

（2）MapView：MapView 是显示地图的 View 组件。它必须和 MapActivity 配合使用，而且只能被 MapActivity 创建，这是因为 MapView 需要通过后台的线程来连接网络或者文件系统，而这些线程需要由 MapActivity 来管理。

（3）MapController：MapController 用于控制地图的移动、缩放等。

（4）OverLay：这是一个可显示于地图之上的可绘制的对象。

（5）GeoPoint：这是一个包含经纬度位置的对象。

10.1.2　导入 Google 地图 API 的 Maps 包

1. 申请 Map API Key 密钥

在进行 Google 地图服务的项目开发之前，必须申请一组验证过的 Map API Key，这样才可以使用 Google 地图服务。Map API Key 的申请过程参见本书的附录 3。

2. 创建 Google API 的 AVD 设备

在创建 AVD 模拟器设备时，选择 Target 项时要选择 Google APIs(Google Inc.)的运行环境项，如图 10.1 所示。

3. 在新建项目时选用 Google API 版本

Google 地图的 API 都集中在 com.google.android.maps 包中，新建项目时，在"New Android Project"对话框中要选择 Google API 版本，如图 10.2 所示。这样，在设计时才能导入 com.google.android.maps 包。

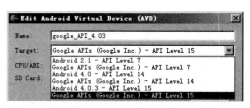
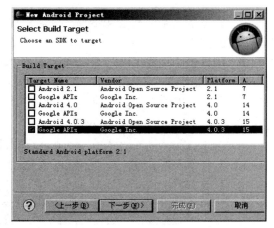

图 10.1　新建 AVD 模拟器设备时选择 Google API 版本　　图 10.2　新建项目时选择 Google API 版本

10.1.3　显示地图 MapView 类

MapView 是 com.google.android.maps 包中显示地图的组件，通常在 MapActivity 中创建 MapView 的对象。MapView 类的常用方法见表 10-2。

表 10-2　MapView 类的常用方法

方　　法	说　　明
displayZoomControls(boolean bl)	设置是否显示缩放控件，参数为 true 或 false
getProjection()	将地图的经度和纬度转换成屏幕像素的实际坐标
getZoomLevel()	返回当前的缩放等级数值
getController()	创建地图控制抽象类 MapController 的对象

续表

方 法	说 明
getMapCenter()	获取地图中心
getLatitudeSpan()	获取纬度值
getLongitudeSpan()	获取经度值
getOverlays()	返回当前所有的 Overlay 层对象
setBuiltInZoomControls(boolean on)	设置是否启用内置缩放控制器
setStreetView()	设置地图显示模式为街道模式
setTraffic()	设置地图显示模式为交通模式
setSatellite()	设置地图显示模式为卫星模式
setZoom()	设置地图缩放率（取值 1~21）

【例 10-1】 创建一个 Google 地图的 View 视图。

创建名为 Ex10_01 的新项目，包名为 com.ex10_01。在新建项目时，要注意选择 Google API 版本。

（1）设计界面布局文件 activity_main.xml。在布局文件 activity_main.xml 中，安排 3 个按钮和一个 Google 地图 View 组件 MapView。3 个按钮分别用于显示"普通地图""交通地图""卫星地图"的视图。MapView 组件需要设置访问 Google 地图的密钥 apiKey 的属性值。界面布局如图 10.3 所示。

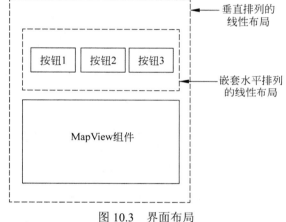

图 10.3 界面布局

完整的 activity_main.xml 的代码如下：

```
1  <?xml version="1.0" encoding="utf-8"?>
2  <LinearLayout
3    xmlns:android="http://schemas.android.com/apk/res/android"
4    android:layout_width="fill_parent"
5    android:layout_height="fill_parent"
6    android:orientation="vertical" >
7    <LinearLayout         ← 嵌套一个线性布局，默认的排列方式（水平排列）
8      android:layout_width="fill_parent"
9      android:layout_height="wrap_content">
10     <Button
11       android:id="@+id/btn1"
12       android:layout_width="100dp"
13       android:layout_height="wrap_content"
14       android:text="普通" />
15     <Button
16       android:id="@+id/btn2"
```

视频演示

```
17            android:layout_width="100dp"
18            android:layout_height="wrap_content"
19            android:text="交通" />
20        <Button
21            android:id="@+id/btn3"
22            android:layout_width="100dp"
23            android:layout_height="wrap_content"
24            android:text="卫星" />
25    </LinearLayout>
26    <com.google.android.maps.MapView
27        android:id="@+id/myMapView1"
28        android:layout_width="fill_parent"
29        android:layout_height="fill_parent"
30        android:layout_x="0px"
31        android:layout_y="82px"
32        android:enabled="true"
33        android:clickable="true"
34        android:apiKey =
35            "0apEt3mipTlINXX7-YkOywzj6i2WmUKTAht_B0A" />      ← Google Map API 密钥,
36        <!-- android:apiKey="Google map API 的密钥" -->          密钥申请参见附录 C
37    </LinearLayout>
```

（2）修改配置文件 AndroidManifest.xml。在 AndroidManifest.xml 文件中添加 com.google.android.maps 元素，由于使用 Google Map API，需要在 AndroidManifest.xml 文件的<application>元素中添加引用 Google Map 库的语句：

```
<uses-library android:name="com.google.android.maps" />
```

在 AndroidManifest.xml 文件中设置访问网络权限，需要在<manifest> 元素中添加取得访问网络权限的语句：

```
<uses-permission android:name="android.permission.INTERNET" />
```

完整的 AndroidManifest.xml 文件的代码如下：

```
1   <?xml version="1.0" encoding="utf-8"?>
2   <manifest xmlns:android="http://schemas.android.com/apk/res/android"
3       package="com.ex10_01"
4       android:versionCode="1"
5       android:versionName="1.0" >
6     <uses-permission android:name =
7           "android.permission.INTERNET" />       ← 允许访问网络的权限
8     <application
9         android:icon="@drawable/ic_launcher"
10        android:label="@string/app_name" >
11      <uses-library android:name =
12           "com.google.android.maps" />          ← 调用 Google Map 库
13      <activity
```

```
14          android:name=".MainActivity"
15          android:label="@string/app_name" >
16          <intent-filter>
17              <action android:name="android.intent.action.MAIN" />
18              <category android:name="android.intent.category.LAUNCHER" />
19          </intent-filter>
20      </activity>
21    </application>
22 </manifest>
```

（3）设计控制文件 MainActivity.java。MainActivity.java 文件是一个显示 Google 地图的 Activity 的子类 MapActivity。

重写 isRouteDisplayed 函数：

```
@Override
protected boolean isRouteDisplayed()
{
   return false;
}
```

设置启动 MapView 内置的缩放功能：

```
mMapView.setBuiltInZoomControls(true);
mMapView.displayZoomControls(true);
```

创建一个监听按钮事件的内部类：

```
class mClick implements OnClickListener
{
    public void onClick(View v)
    {
      if(v == dtBtn){          //单击 "普通" 按钮时触发
          mMapView.setTraffic(false);
          mMapView.setSatellite(false);
      }
      else if(v == jtBtn){     //单击 "交通" 按钮时触发
          mMapView.setTraffic(true);
          mMapView.setSatellite(false);
      }
      else if(v == wxBtn) {    //单击 "卫星" 按钮时触发
          mMapView.setSatellite(true);
          mMapView.setTraffic(false);
      }
    }
}
```

完整程序如下：

```
1   package com.ex10_01;
2   import com.google.android.maps.MapActivity;
3   import com.google.android.maps.MapView;
```

```
4    import android.os.Bundle;
5    import android.view.View;
6    import android.view.View.OnClickListener;
7    import android.widget.Button;
8
9    public class MainActivity extends MapActivity
10   {
11       MapView mMapView;
12       Button dtBtn, jtBtn, wxBtn;
13       @Override
14       public void onCreate(Bundle savedInstanceState)
15       {
16           super.onCreate(savedInstanceState);
17           setContentView(R.layout.main);
18           mMapView=(MapView)findViewById(R.id.myMapView1);
19           mMapView.setBuiltInZoomControls(true);         ┐← 设置地图的缩放功能
20           mMapView.displayZoomControls(true);            ┘
21           dtBtn=(Button)findViewById(R.id.btn1);
22           jtBtn=(Button)findViewById(R.id.btn2);
23           wxBtn=(Button)findViewById(R.id.btn3);
24           dtBtn.setOnClickListener(new mClick());
25           jtBtn.setOnClickListener(new mClick());
26           wxBtn.setOnClickListener(new mClick());
27       }
28       class mClick implements OnClickListener     ← 监听按钮事件
29       {
30         @Override
31         public void onClick(View v)
32         {
33          if(v==dtBtn){
34              mMapView.setTraffic(false);           ┐← 显示普通地图时,交通地图
35              mMapView.setSatellite(false);         ┘   和卫星地图功能失效
36          }
37          else if(v==jtBtn){
38              mMapView.setTraffic(true);            ┐← 单击"交通"按钮时,交通地
39              mMapView.setSatellite(false);         ┘   图有效,但卫星地图功能失效
40          }
41          else if(v==wxBtn) {
42              mMapView.setSatellite(true);          ┐← 单击"卫星"按钮时,卫星地
43              mMapView.setTraffic(false);           ┘   图有效,但交通地图功能失效
44          }
45         }
46       }
47       @Override
48       protected boolean isRouteDisplayed()
```

```
49      {   //TODO Auto-generated method stub
50          return false;
51      }
52  }
```

程序的运行结果如图 10.4 所示。

图 10.4　显示 Google 地图 MapView

10.1.4　添加 Google 地图的贴图

在地图应用程序中，有时只有地图不够，还要在地图上添加一些标注，这时需要在地图上建立贴图。地图和地图贴图分别位于两个不同的图层。

1. 贴图类 Overlay

在 Android 系统中要建立地图的贴图，需要创建贴图类 Overlay 的子类。Overlay 常用方法见表 10-3。

表 10-3　贴图类 Overlay 的常用方法

方　　法	说　　明
draw(Canvas canvas, MapView mapView, boolean shadow, long when)	在地图贴片图层上绘制标注
drawAt(android.graphics.Canvas canvas, android.graphics.drawable.Drawable drawable, int x, int y, boolean shadow)	在指定坐标（x, y）处绘制标注
onKeyDown(int keyCode, android.view.KeyEvent event, MapView mapView)	处理按下某个按键事件
onKeyUp(int keyCode, android.view.KeyEvent event, MapView mapView)	处理抬起某个按键事件
onTouchEvent(android.view.MotionEvent e, MapView mapView)	处理触摸屏事件

2. 经纬度位置类 GeoPoint

GeoPoint 是表示一组经度和纬度位置的类，其常用方法见表 10-4。

表 10-4 经纬度位置类 GeoPoint 的常用方法

方 法	说 明
GeoPoint(int latitudeE6, int longitudeE6)	创建指定经纬度位置对象的构造方法
getLatitudeE6()	获取 GeoPoint 对象的纬度值
getLongitudeE6()	获取 GeoPoint 对象的经度值

构造方法 GeoPoint(int latitudeE6, int longitudeE6)的参数分别为经度和纬度，以微度为单位（度*1E6）。纬度 latitudeE6 的取值范围为-80～80，经度 longitudeE6 的取值范围为-180～180。

3. 将经纬度转换成实际屏幕坐标

为了确定贴图的显示位置，需要将经纬度转换成实际屏幕坐标：

```
Point screenPoint = new Point();         //屏幕坐标对象
GeoPoint geoPoint;                       //经纬度对象
geoPoint = new GeoPoint((int)(24.369647 * 1E6), (int)(118.043226 * 1E6));
mapView.getProjection().toPixels(geoPoint, screenPoint);
```

【例 10-2】在 Google 地图 View 上添加贴图标注。

新建项目 ex10_02，并将事先准备的图片 point.jpg 复制到资源 res\drawable-hdpi 目录下。

设计控制文件 MainActivity.java，增加一个地图贴图 Overlay 的子类 MyOverlay。

视频演示

（1）定义经纬度位置对象并转换为屏幕像素坐标：

```
GeoPoint geoPoint=new GeoPoint((int)(24.369647 * 1E6),
                               (int)(118.043226 * 1E6));
Point screenPoint=new Point();
mapView.getProjection().toPixels(geoPoint, screenPoint);
```

（2）在 draw()方法中绘制贴图标注的图片和文字：

```
canvas.drawBitmap(bmp, screenPoint.x, screenPoint.y, paint);
canvas.drawText("旅游目的地", screenPoint.x, screenPoint.y, paint);
```

（3）在列表中添加贴图对象 Overlay，显示标注信息：

```
MyOverlay mOverlay=new MyOverlay();
List<Overlay> list=mMapView.getOverlays();
list.add(mOverlay);
```

程序如下：

```
1   package com.ex10_02;
2   import java.util.List;
3   import com.google.android.maps.GeoPoint;
4   import com.google.android.maps.MapActivity;
5   import com.google.android.maps.MapView;
```

```java
6   import com.google.android.maps.Overlay;
7   import android.graphics.Bitmap;
8   import android.graphics.BitmapFactory;
9   import android.graphics.Canvas;
10  import android.graphics.Paint;
11  import android.graphics.Point;
12  import android.os.Bundle;
13
14  public class MainActivity extends MapActivity
15  {
16    MapView mMapView;
17    @Override
18    public void onCreate(Bundle savedInstanceState)
19    {
20      super.onCreate(savedInstanceState);
21      setContentView(R.layout.main);
22      mMapView=(MapView)findViewById(R.id.myMapView1);
23      mMapView.setBuiltInZoomControls(true);
24      mMapView.displayZoomControls(true);
25      /*添加Overlay,用于显示标注信息*/
26      MyOverlay mOverlay=new MyOverlay();
27      List<Overlay> list=mMapView.getOverlays();
28      list.add(mOverlay);
29    }
30  /*在地图上贴图*/
31  class MyOverlay extends Overlay
32  {
33   public boolean draw(Canvas canvas, MapView mapView,
34                       boolean shadow, long when)
35    {
36      super.draw(canvas, mapView, shadow);
37      Paint paint=new Paint();
38      Point screenPoint=new Point();
39      /*将经纬度转换成实际屏幕像素坐标*/
40      GeoPoint geoPoint=new GeoPoint((int)(24.369647 * 1E6),
41                                    (int)(118.043226 * 1E6) );
42      mapView.getProjection().toPixels(geoPoint, screenPoint);
43      paint.setStrokeWidth(1);
44      paint.setARGB(255, 0, 0, 255);
45      paint.setStyle(Paint.Style.STROKE);
46      paint.setTextSize(14);
47      Bitmap bmp=BitmapFactory.decodeResource(getResources(),
48                          R.drawable.point);
49      /*在地图的贴片图层上绘制图片*/
50      canvas.drawBitmap(bmp, screenPoint.x, screenPoint.y, paint);
```

```
51          /*在地图的贴片图层上绘制文字*/
52          canvas.drawText("旅游目的地", screenPoint.x, screenPoint.y, paint);
53          return true;
54        }
55    }
56    @Override
57    protected boolean isRouteDisplayed()     ← MapActivity 的方法，必须实现
58    {
59        return false;
60    }
61 }
```

用户界面程序的 Google map API 密钥设置及修改配置文件的访问权限同例 10-1，在此不再赘述。

程序的运行结果如图 10.5 所示。

图 10.5　在地图上添加贴图标志

10.2　位置服务

所谓位置服务，即确定某点地理位置的经纬度。在 Android 系统中，由定位服务的系统包 android.location 提供位置服务。

1. 定位系统包 location

定位系统包 location 中主要有以下几类：

- LocationManager 类是位置服务的核心组件，提供访问定位服务的功能，它提供了一系列方法来处理与地理位置有关的问题。另外，临近警报功能也可以借助该类来实现。其主要方法见表 10-5。

- LocationProvider 类是定位提供者的抽象类。定位提供者具备周期性报告设备地理位置的功能。
- LocationListener 类提供定位信息发生改变时的回调功能，但必须事先在定位管理器中注册监听器对象。
- Criteria 类使得应用能够通过在 LocationProvider 中设置的属性来选择合适的定位提供者。
- Location 类为位置抽象类，通过它可以获得相关的位置数据。其主要方法见表 10-6。

表 10-5 LocationManager 类的常用方法

方　　法	说　　明
requestLocationUpdates (String provider, 　　　　　　long minTime, 　　　　　　float minDistance, 　　　　　　LocationListener listener)	注册一个周期性的更新位置的方法
getLastKnownLocation(String provider)	获取最近一次的位置信息
getBestProvider(Criteria criteria, 　　boolean enabledOnly)	获取 Criteria 对象中设定的设备，如 GPS

注册位置方法 requestLocationUpdates()有 4 个参数，其含义如下：
- provider 为位置提供设备的名称；
- minTime 为时间间隔，以毫秒为单位；
- minDistance 为最小距离间隔，以米为单位；
- listener 为位置更新的监听接口对象，必须实现它的 4 个方法。

需要指出的是，LocationManager 类不是通过构造方法构造对象，而是通过建立对象引用来获取：

```
LocationManager locatmanager=
  (LocationManager)getSystemService(Context.LOCATION_SERVICE);
```

表 10-6 Location 类的常用方法

方　　法	说　　明
getLatitude()	获取纬度
getLongitude()	获取经度
getAltitude()	获取海拔

2. 实现地理位置定位的主要步骤

1）创建 LocationManager 对象

```
LocationManager locatmanager=
    (LocationManager)getSystemService(Context.LOCATION_SERVICE);
```

2）获取 GPS 设备

```
String providerName=
    locatmanager.getBestProvider(getCriteria(), true);
```

3）获取位置信息

```
Location location=
    locatmanager.getLastKnownLocation(providerName);
```

4）注册 LocationListener 接口对象的监听器

```
mLocationListener listen=new mLocationListener();
locatmanager.requestLocationUpdates(providerName, 5000, 8, listen);
```

实现 LocationListener 接口时，必须覆盖它的 4 个方法：

- public void onLocationChanged(Location location);
- public void onProviderDisabled(String provider);
- public void onProviderEnabled(String provider);
- public void onStatusChanged(String provider, int status, Bundle extras);

3. 应用示例

【例 10-3】确定当前所处地理位置示例。

程序代码如下：

```
1   package com.example.ex10_03;
2   import android.app.Activity;
3   import android.content.Context;
4   import android.location.Criteria;
5   import android.location.Location;
6   import android.location.LocationListener;
7   import android.location.LocationManager;
8   import android.os.Bundle;
9   import android.widget.EditText;
10
11  public class MainActivity extends Activity
12  {
13      LocationManager locatmanager;        ← 声明 LocationManager 对象
14      EditText edit;
15      @Override
16      public void onCreate(Bundle savedInstanceState)
17      {
18          super.onCreate(savedInstanceState);
19          setContentView(R.layout.main);
20          edit=(EditText)findViewById(R.id.editText1);
21          locatmanager=                    ← 创建 LocationManager 对象
22              (LocationManager)getSystemService(Context.LOCATION_SERVICE);
23          String providerName=
24              locatmanager.getBestProvider(getCriteria(), true);   ← 获取 GPS 设备
25          Location location=
26              locatmanager.getLastKnownLocation(providerName);     ← 获取位置信息
27          /** 添加 LocationListener 监听器
28           * 注册一个周期性的位置更新：需要从 GPS 获取位置信息，
29           * 每隔 2000ms 更新一次，最小间隔距离为 5m
```

```
30         */
31         mLocationListener listen=new mLocationListener();
32         locatmanager.requestLocationUpdates(providerName, 2000, 5, listen);
33         updateText(location);    ← 更新 EditText 控件的位置数据
34     }
35     /*地理位置查询设置*/
36     public Criteria getCriteria()
37     {
38       Criteria criteria=new Criteria();
39       criteria.setAccuracy(Criteria.ACCURACY_COARSE);  ← 设置查询精度
40       criteria.setSpeedRequired(false);   ← 设置是否要求速度
41       criteria.setCostAllowed(false);     ← 设置是否允许产生费用
42       criteria.setBearingRequired(false); ← 设置是否需要得到方向
43       criteria.setAltitudeRequired(false);← 设置是否要得到海拔高度
44       return criteria;                    ← 返回查询设置
45     }
46     /*在 EditText 中显示位置数据*/
47     public void updateText(Location newLocation)
48     {
49       if(location !=null)  ← 判断是否为空
50       {
51           double latitude=location.getLatitude();   ← 获取纬度
52           double longitude=location.getLongitude(); ← 获取经度
53           edit.setText("您现在的位置是\n 纬度: ");
54           edit.append(String.valueOf(latitude));    ← 显示纬度
55           edit.append("\n 经度: ");
56           edit.append(String.valueOf(longitude));   ← 显示经度
57       }
58       else{
59           edit.getEditableText().clear();   ← 如果传入的 Location 对象为空则清空
60       }
61     }
62
63     class mLocationListener implements LocationListener
64     {
65       @Override
66       public void onLocationChanged(Location location)
67       {//当坐标改变时触发此函数,如果 Provider 传进相同的坐标,它就不会被触发
68           updateText(location);
69       }
70       @Override
71       public void onProviderDisabled(String provider)
72       {    //Provider 被 disable 时触发此函数,例如 GPS 被关闭
73           updateText(null);
74       }
```

```
75      @Override
76      public void onProviderEnabled(String provider)
77      {    //Provider 被 enable 时触发此函数，例如 GPS 被打开
78          Location locat;
79          locat=locatmanager.getLastKnownLocation(provider);    ← 获取位置信息
80          updateText(locat);    ← 更新显示的位置数据
81      }
82      @Override
83      public void onStatusChanged(String provider, int status, Bundle extras)
84      {    //Provider 改变状态时触发此函数
85      }
86  }
87 }
```

修改配置文件 AndroidManifest.xml，要设置使用 GPS 的权限及访问网络的权限：

```
<uses-permission
 android:name="android.permission.ACCESS_COARSE_LOCATION" />
<uses-permission
 android:name="android.permission.ACCESS_FINE_LOCATION" />
<uses-permission android:name="android.permission.INTERNET" />
```

在模拟器中测试程序时，由于模拟器无法获取地理位置，可以设置一个模拟的坐标值。打开 DDMS 工具，选择"窗口"（Window）→"显示视图"（Show View）→Emulator Control 命令，即可看到如图 10.6（a）所示的对话框，手动输入一个模拟当前位置的经纬度后，运行程序，结果如图 10.6（b）所示。

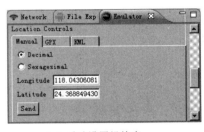
（a）手动设置经纬度

（b）在模拟器上显示当前位置

图 10.6 显示当前位置

10.3 传感器检测技术

10.3.1 传感器简介

传感器（Sensor）是一种检测装置，它能检测和感受到外界的信号，并将信息转换成电信号或其他所需形式的信息输出，以满足信息的传输、处理、存储、显示、记录和控制等要求。

1. 传感器的类型

Android 系统中内置了很多类型的传感器，这些传感器被封装在 Sensor 类中。Sensor 是管理各种传感器的共同属性（名称、版本等）的类，Sensor 类包含一个常量集合，用于描述 Sensor 对象所表示的硬件传感器类型，这些常量均以 Sensor.TYPE_<TYPE>的形式表示。Android 系统的常见传感器类型见表 10-7。

表 10-7 Android 系统常见的传感器类型

类 型 常 量	说 明
Sensor.TYPE_ACCELEROMETER	加速度（重力）传感器
Sensor.TYPE_LIGHT	光线传感器
Sensor.TYPE_MAGNETIC_FIELD	磁场传感器
Sensor.TYPE_PROXIMITY	距离（临近性）传感器
Sensor.TYPE_AMBIENT_TEMPERATURE	温度传感器
Sensor.TYPE_PRESSURE	压力传感器
Sensor.TYPE_ALL	所有类型的传感器

2. 与传感器相关的类

除 Sensor 类外，要使用传感器，还需要用到传感器管理类 SensorManager 和传感器事件监听接口 SensorEventListener。

1）传感器管理类 SensorManager

Android 中的所有传感器都需要通过 SensorManager 对象来访问，SensorManager 没有构造方法，需要调用 getSystemService(SENSOR_SERVICE)方法创建传感器管理对象。

```
SensorManager mSensorMgr = (SensorManager) getSystemService(SENSOR_SERVICE);
```

SensorManager 类的常用方法见表 10-8。

表 10-8 传感器管理类 SensorManager 的常用方法

方 法	说 明
getSensorList(int type)	获取传感器类型列表
registerListener(SensorEventListener listener,Sensor sensor,int rate)	注册传感器的监听器
unregisterListener(SensorEventListener listener)	注销传感器的监听器
getDefaultSensor(int type)	获取默认的传感器对象

在 registerListener()方法中，第 3 个参数 rate 为传感器的更新速率，其更新速率分为 4 个级别：

- SensorManager.SENSOR_DELAY_FASTEST 为最快级，特别敏感，一般不推荐使用。
- SensorManager.SENSOR_DELAY_GAME 为游戏级，实时性较高的游戏使用。
- SensorManager.SENSOR_DELAY_NORMAL 为普通级，默认使用。
- SensorManager.SENSOR_DELAY_UI 为用户界面级，一般屏幕更新使用。

2）实现 SensorEventListener 接口

传感器事件监听接口 SensorEventListener 有两个方法必须实现。
- onAccuracyChanged(Sensor sensor,int accuracy)：传感器的精度变化时，此方法被调用。
- onSensorChanged(SensorEvent event)：传感器的值改变时，此方法被调用。

【例 10-4】列出传感器的类型。

程序代码如下：

```
1   package com.example.sensortest;
2   import java.util.List;
3   import android.hardware.Sensor;
4   import android.hardware.SensorEvent;
5   import android.hardware.SensorEventListener;
6   import android.hardware.SensorManager;
7   import android.os.Bundle;
8   import android.widget.LinearLayout;
9   import android.widget.TextView;
10  import android.app.Activity;
11
12  public class MainActivity extends Activity
13   implements SensorEventListener
14  {
15    private SensorManager sensorManager;
16    @Override
17    public void onCreate(Bundle savedInstanceState)
18    {
19        super.onCreate(savedInstanceState);
20        setContentView(R.layout.activity_main);
21    }
22    @Override
23    protected void onResume()
24    {
25        super.onResume();
26        sensorManager=
27          (SensorManager)this.getSystemService(SENSOR_SERVICE);    ← 获取传感器管理服务
28        List<Sensor> sensorList=
29          sensorManager.getSensorList(Sensor.TYPE_ALL);
30        LinearLayout layout=new LinearLayout(this);    ← 创建布局对象
31        layout.setOrientation(LinearLayout.VERTICAL);
32        TextView txt;
33        for(Sensor s:sensorList)
34        {
35            txt=new TextView(this);
36            txt.setText(s.getName());
```

```
37              layout.addView(txt,new LinearLayout.LayoutParams(
38                      LinearLayout.LayoutParams.FILL_PARENT,
39                      LinearLayout.LayoutParams.WRAP_CONTENT));
40          }
41          setContentView(layout);
42      }
43      @Override
44      public void onAccuracyChanged(Sensor sensor, int accuracy)
45      {           }
46      @Override
47      public void onSensorChanged(SensorEvent event)
48      {           }
49  }
```
设置布局

在真实手机上运行程序，列出传感器的类型，如图 10.7 所示。

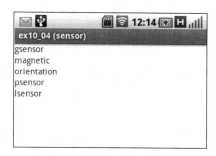

图 10.7　列出传感器的类型

10.3.2　加速度传感器的应用示例

加速度传感器是用于检测物体加速度的传感器。物体在运动时其加速度也跟着变化，如果能获取到加速度的值，就可以知道物体受到什么样的作用力或物体进行什么样的运动。

通过 Android 的加速度传感器可以从 X、Y、Z 3 个方向轴获取加速度。X、Y、Z 3 个方向轴的定义为：

- X 轴的方向是沿着手机屏幕从左向右的方向。
- Y 轴的方向是从手机屏幕的左下角开始沿着屏幕的上下方向指向屏幕的顶端。
- Z 轴的方向是从手机里指向外的前后方向。

加速度传感器的方向轴如图 10.8 所示。

图 10.8　加速度传感器的方向轴

通过 SensorEventListener 接口的 onSensorChanged(SensorEvent event)可以获取 X、Y、Z 3 个方向轴重力加速度的值。

例如，设 X 轴、Y 轴、Z 轴 3 个方向的重力分量的值分别为 x、y、z，则：

```
public void onSensorChanged(SensorEvent event)
{
    float x=event.values[0];
    float y=event.values[1];
    float z=event.values[2];
}
```

【例10-5】将手机设置成振动状态,摇一摇后,立刻停止振动。

本例要解决两个问题:(1)将手机设为振动状态;(2)应用加速度传感器的工作原理,快速晃动手机,即摇一摇后,使手机的振动停止。

Android 手机的振动控制由 Vibrator 类实现。Vibrator 类主要有两种方法。

(1)vibrate(long[] pattern, int repeat)方法:设置振动周期。

(2)cancel()方法:停止振动。

设置振动事件,需要知道振动的时间长短、振动的周期等。在 Android 中,振动的时间以毫秒为单位(1/1000 秒),注意,如果设置的时间值太小,会感觉不出来。

通过调用 Vibrator 类的 vibrate(long[] pattern, int repeat)方法来实现振动功能。Vibrate() 有两个参数:

- 第 1 个参数 long[] pattern:为设置振动效果的数组,数组中数字的含义依次为[静止时长,振动时长,静止时长,振动时长],时长的单位是毫秒。
- 第 2 个参数 repeat:取值-1 表示只振动一次,取值 0 表示振动会一直持续。

需要指出的是,Vibrator 类没有构造方法,需要通过建立对象引用的 getSystemService() 方法获取 Vibrator 对象:

```
SensorManager mSensorManager=
    (SensorManager)getSystemService(SENSOR_SERVICE);
```

若要停止振动,调用 Vibrator 类的 cancel()方法即可。

程序代码如下:

```
1   package com.ex10_05;
2   import android.app.Activity;
3   import android.app.Service;
4   import android.hardware.Sensor;
5   import android.hardware.SensorEvent;
6   import android.hardware.SensorEventListener;
7   import android.hardware.SensorManager;
8   import android.os.Bundle;
9   import android.os.Vibrator;
10  import android.view.View;
11  import android.view.View.OnClickListener;
12  import android.widget.Button;
13
14  public class MainActivity extends Activity implements SensorEventListener
15  {
```

视频演示

```
16      Button clearBtn;
17      private SensorManager mSensorManager;
18      private Vibrator vibrator;          ◄── 声明振动对象
19      @Override
20      public void onCreate(Bundle savedInstanceState)
21      {
22          super.onCreate(savedInstanceState);
23          setContentView(R.layout.main);
24          mSensorManager=
25              (SensorManager)getSystemService(SENSOR_SERVICE);   ◄── 获取传感器管理服务
26          vibrator=(Vibrator)getApplication()
27                  .getSystemService(Service.VIBRATOR_SERVICE);   ◄── 获取振动对象
28          clearBtn=(Button)findViewById(R.id.clear);
29          clearBtn.setOnClickListener(new mClick());
30      }
31
32  class mClick  implements OnClickListener
33  {
34     public void onClick(View v)
35     {
36      clearBtn.setText("改振动喽~");
37      try{
38          vibrator.vibrate(new long[]{100, 100, 100, 1000}, 0);   ◄── 设置振动周期
39        }catch(Exception e){
40          System.out.println("振动错误！！！！！！");
41        }
42     }
43  }
44  @Override
45  protected void onResume()
46  {
47    super.onResume();
48    mSensorManager.registerListener(this,                ◄── 为加速度传感器注册监听器
49      mSensorManager.getDefaultSensor(Sensor.TYPE_ACCELEROMETER),
50      SensorManager.SENSOR_DELAY_NORMAL );
51  }
52  @Override
53  protected void onStop()
54  {
55          super.onStop();
56  }
57  @Override
58  protected void onPause()
59  {
60      super.onPause();
```

```
61      }
62      public void onAccuracyChanged(Sensor sensor, int accuracy)
63      {           }
64      public void onSensorChanged(SensorEvent event)   ← 当传感器的值发生
65      {                                                   改变时回调该方法
66        int sensorType=event.sensor.getType();
67        float[] values=event.values;           ← values[0]:X 轴，values[1]: Y
68        if(sensorType==Sensor.TYPE_ACCELEROMETER )         轴，values[2]: Z 轴
69        {
70          /**由于正常情况下，任意轴数值在 9.8~10 之间，
71           * 当突然摇动手机的时候，瞬时加速度突然增大或减少，
72           * 所以，只需监听任意轴的加速度大于 14，改变需要的设置即可
73           */
74          if((Math.abs(values[0])>14 || Math.abs(values[1])>14
75                              || Math.abs(values[2])>14))
76          {
77              clearBtn.setText("别摇了，头晕死了，已经停止振动");
78              vibrator.cancel();  ← 摇动手机后，停止振动
79          }
80        }
81      }
82      public void onAccuracyChanged(int sensor, int accuracy)
83      {           }
84      public void onSensorChanged(int sensor, float[] values)
85      {           }
86  }
```

修改配置文件 AndroidManifest.xml，设置允许使用振动效果的权限：

```
<uses-permission android: name = "android.permission.VIBRATE"/>
```

由于模拟器无法实现振动功能，只能在真实手机上运行。程序运行结果如图 10.9 所示。当单击"设为振动"按钮后，手机产生振动，快速摇动手机，振动停止。

【例 10-6】简单的重力小球游戏。

程序代码如下：

```
1   package com.example.ex10_06;
2   import android.app.Activity;
3   import android.content.Context;
4   import android.content.pm.ActivityInfo;
5   import android.graphics.Bitmap;
6   import android.graphics.BitmapFactory;
7   import android.graphics.Canvas;
8   import android.graphics.Color;
9   import android.graphics.Paint;
10  import android.hardware.Sensor;
11  import android.hardware.SensorEvent;
```

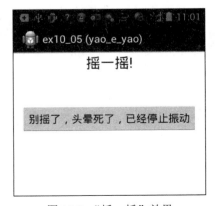

图 10.9 "摇一摇"效果

```java
12  import android.hardware.SensorEventListener;
13  import android.hardware.SensorManager;
14  import android.os.Bundle;
15  import android.view.SurfaceHolder;
16  import android.view.SurfaceView;
17  import android.view.Window;
18  import android.view.WindowManager;
19  import android.view.SurfaceHolder.Callback;
20  public class MainActivity extends Activity
21  {
22    BallView mAnimView = null;
23    @Override
24    public void onCreate(Bundle savedInstanceState)
25    {
26      super.onCreate(savedInstanceState);
27      /*全屏显示窗口*/
28      requestWindowFeature(Window.FEATURE_NO_TITLE);
29      getWindow().setFlags(WindowManager.LayoutParams.FLAG_FULLSCREEN,
30      WindowManager.LayoutParams.FLAG_FULLSCREEN);
31      /*强制横屏*/
32      setRequestedOrientation(ActivityInfo.SCREEN_ORIENTATION_LANDSCAPE);
33      /*显示自定义的游戏View*/
34      mAnimView = new BallView(this);
35      setContentView(mAnimView);
36    }
37
38    public class BallView extends SurfaceView
39          implements Callback,Runnable,SensorEventListener
40    {
41      /*每50帧刷新一次屏幕*/
42      public static final int TIME_IN_FRAME=50;
43      /*游戏画笔*/
44      Paint mPaint=null;
45      Paint mTextPaint=null;
46      SurfaceHolder mSurfaceHolder=null;
47      /*控制游戏更新循环*/
48      boolean mRunning=false;
49      /*游戏画布*/
50      Canvas mCanvas=null;
51      /*控制游戏循环*/
52      boolean mIsRunning=false;
53      /*SensorManager 管理器*/
54      private SensorManager mSensorMgr=null;
55      Sensor mSensor=null;
56      /*手机屏幕宽高*/
```

```
57      int mScreenWidth=0;
58      int mScreenHeight=0;
59      /*小球资源文件越界区域*/
60      private int mScreenBallWidth=0;
61      private int mScreenBallHeight=0;
62      /*游戏背景文件*/
63      private Bitmap mbitmapBg;
64      /*小球资源文件*/
65      private Bitmap mbitmapBall;
66      /*小球的坐标位置*/
67      private float mPosX=200;
68      private float mPosY=0;
69      /*重力感应X轴、Y轴、Z轴的重力值*/
70      private float mGX=0;
71      private float mGY=0;
72      private float mGZ=0;
73      public BallView(Context context)
74      {
75        super(context);
76        /*设置当前View拥有控制焦点*/
77        this.setFocusable(true);
78        /*设置当前View拥有触摸事件*/
79        this.setFocusableInTouchMode(true);
80        /*获取SurfaceHolder对象*/
81        mSurfaceHolder = this.getHolder();
82        /*将mSurfaceHolder添加到Callback回调函数中*/
83        mSurfaceHolder.addCallback(this);
84        /*创建画布*/
85        mCanvas = new Canvas();
86        /*创建曲线画笔*/
87        mPaint = new Paint();
88        mPaint.setColor(Color.WHITE);
89        /*加载小球资源*/
90        mbitmapBall = BitmapFactory.decodeResource(this.getResources(),
91                               R.drawable.ball);
92        /*加载游戏背景*/
93        mbitmapBg = BitmapFactory.decodeResource(this.getResources(),
94                               R.drawable.bg);
95        /*获取SensorManager对象*/
96        mSensorMgr = (SensorManager) getSystemService(SENSOR_SERVICE);
97        mSensor = mSensorMgr.getDefaultSensor(Sensor.TYPE_ACCELEROMETER);
98        mSensorMgr.registerListener(this, mSensor,
99                               SensorManager.SENSOR_DELAY_GAME);
100     }
101     private void Draw()
```

```
102     {
103         /*绘制游戏背景*/
104         mCanvas.drawBitmap(mbitmapBg,0,0, mPaint);
105         /*绘制小球*/
106         mCanvas.drawBitmap(mbitmapBall, mPosX,mPosY, mPaint);
107         /*显示X轴、Y轴、Z轴的重力值*/
108         mCanvas.drawText("   X轴重力值: " + mGX, 0, 20, mPaint);
109         mCanvas.drawText("   Y轴重力值: " + mGY, 0, 40, mPaint);
110         mCanvas.drawText("   Z轴重力值: " + mGZ, 0, 60, mPaint);
111     }
112     @Override
113     public void surfaceChanged(SurfaceHolder holder, int format,
114                                 int width,int height) {     }
115     @Override
116     public void surfaceCreated(SurfaceHolder holder)
117     {
118         /*开始游戏主循环线程*/
119         mIsRunning=true;
120         new Thread(this).start();
121         /*得到当前屏幕的宽高*/
122         mScreenWidth=this.getWidth();
123         mScreenHeight=this.getHeight();
124         /*得到小球越界区域*/
125         mScreenBallWidth=mScreenWidth - mbitmapBall.getWidth();
126         mScreenBallHeight=mScreenHeight - mbitmapBall.getHeight();
127     }
128     @Override
129     public void surfaceDestroyed(SurfaceHolder holder)
130     {
131         mIsRunning=false;
132     }
133     @Override
134     public void run()
135     {
136       while (mIsRunning)
137       {
138         /*取得更新游戏之前的时间*/
139         long startTime=System.currentTimeMillis();
140         /*在这里加上线程安全锁*/
141         synchronized (mSurfaceHolder)
142         {
143             /*拿到当前画布,然后锁定*/
144             mCanvas=mSurfaceHolder.lockCanvas();
145             Draw();
146             /*绘制结束后解锁显示在屏幕上*/
```

```
147        mSurfaceHolder.unlockCanvasAndPost(mCanvas);
148      }
149      /*取得更新游戏结束的时间*/
150      long endTime=System.currentTimeMillis();
151      /*计算出游戏一次更新的毫秒数*/
152      int diffTime=(int)(endTime-startTime);
153      /*确保每次更新时间为50帧*/
154      while(diffTime<=TIME_IN_FRAME)
155      {
156        diffTime=(int)(System.currentTimeMillis()-startTime);
157        /*线程等待*/
158        Thread.yield();
159      }
160    }
161 }
162 @Override
163 public void onAccuracyChanged(Sensor arg0, int arg1)
164 { }
165 @Override
166 public void onSensorChanged(SensorEvent event)
167 {
168   mGX=event.values[0];           ← 获取 X、Y、Z 3 个方向上重力加速度的改变值
169   mGY=event.values[1];
170   mGZ=event.values[2];
171   mPosX += mGX * 2;              ← 这里乘以 2 是为了让小球移动的更快
172   mPosY += mGY * 2;
173   if (mPosX<0)
174   {
175      mPosX=0;
176   } else if(mPosX > mScreenBallWidth)
177   {
178     mPosX=mScreenBallWidth;
179   }
180   if (mPosY<0)                   ← 检测小球是否超出边界
181   {
182     mPosY=0;
183   } else if(mPosY > mScreenBallHeight)
184   {
185     mPosY=mScreenBallHeight;
186   }
187 }
188 }
189 }
```

程序在真实手机上运行的结果如图 10.10 所示。

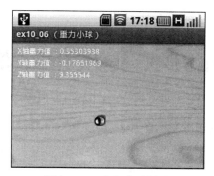

图 10.10 重力小球游戏

习 题 10

1．编写一个地图应用程序,设已知 3 个地点位置的经纬度分别为（24.477384, 118.158216）、（24.488967, 118.144277）、（24.491091, 118.136781）, 在地图上画出它们的路线。

提示：
设
```
int x1, y1, x2, y2, x3, y3;
x1=(int)(24.477384 * 1000000);
y1=(int)(118.158216 * 1000000);
x2=(int)(24.488967 * 1000000);
y2=(int)(118.144277 * 1000000);
x3=(int)(24.491091 * 1000000);
y3=(int)(118.136781 * 1000000);
```
则
```
GeoPoint gpoint1, gpoint2, gpoint3;   // 连线的点
gpoint1=new GeoPoint(x1,y1);
gpoint2=new GeoPoint(x2,y2);
gpoint3=new GeoPoint(x3,y3);
mMapView.getController().animateTo(gpoint1);
```

2．编写一个通过摇一摇播放音乐的音频播放器。

3．进一步完善例 10-5, 添加开始、结束等功能, 使之成为一个较完整的简单游戏。

附录 A JavaSDK 及 Eclipse 的安装与配置

1. Java SDK 的安装与环境变量配置

1）下载与安装

在 Oracle 公司的官方网站 http://www.oracle.com/technetwork/java/javase/downloads/ 下载最新版本系统软件 Java SDK（简称 JDK）。对于使用 Windows 操作系统的用户来说，一般下载 Windows x86 版：jdk-7u7-windows-i586.exe。

下载了必要的软件之后，即可按文档 README.TXT 介绍的安装方法安装和配置 Java 的开发环境了。

在 Windows 环境下，直接单击下载 JDK 安装文件的图标，即可自动进入安装过程，此时可以按照提示过程，逐步完成安装。

2）Java 开发环境的配置

安装 JDK 后，还需要配置 Java 的运行环境变量。Java SDK 中有两个相关环境变量，即 CLASSPATH 和 PATH，它们分别指定了 Java 的类路径和 JDK 命令搜索路径。在这里假设 JDK 安装在"C:\Java\JDK7"目录下。

在 Windows XP 下，环境变量的设置可以在桌面上右击"我的电脑"，在弹出的"系统属性"对话框中选择"高级"选项卡，单击"环境变量"按钮，然后在弹出的"编辑系统变量"对话框中进行设置。如在"变量名"文本框中输入 classpath，在"变量值"文本框中输入"C:\Java\jdk7\lib\dt.jar；C:\Java\JDK7\lib\tools.jar"，最后单击"确定"按钮。

用相同的方法，建立变量名 Path，其变量值为"C:\Java\jdk7\bin"，如附图 A.1 所示。

（a）设置环境变量 classpath　　　　　　　　（b）设置环境变量 Path

附图 A.1　设置环境变量

经过开发环境配置以后，无论当前目录是在何处，执行诸如 Javac、Java 等命令时，操作系统都会找到这些文件并执行它们，从而使用户可以在任何目录下执行 Java 程序代码。

2. Eclipse 的安装与配置

1）下载与安装

Eclipse 是一个开放源代码的、基于 Java 的可扩展开发平台。读者可以到其网站"http://www.

eclipse.org/downloads/"下载最新版本系统软件 Eclipse IDE for Java Developers。选择 Windows 32 Bit 版，如附图 A.2 所示。

附图 A.2 下载 Eclipse

安装 Eclipse 很简单，将其解压到指定目录即可。

2）Eclipse 的配置

Eclipse 一般不需要配置，指定工作目录直接使用即可。

附录 B Android 的调试工具

1. Android 的 ADB 工具使用

在 Android SDK 的 Tools 目录下包含 Android 模拟器操作的重要命令 ADB，ADB 的全称为 Android Debug Bridge，借助这个工具，用户可以管理设备或手机模拟器的状态，还可以在设备上运行 Shell 命令的操作。

默认情况下，当运行 Eclipse 时 ADB 进程会自动运行。

下面是 ADB 的一些常用操作命令。

1）查看 ADB 的版本信息

```
adb version
```

例如，打开 Windows 的命令行窗口，输入命令 adb version：

```
C:\Documents and Settings\Administrator>adb version
Android Debug Bridge version 1.0.29
```

2）安装应用到模拟器

```
adb install [-l] [-r] <file>
```

其中，file 是需要安装的 apk 文件的绝对路径。

3）进入设备或模拟器的 shell 环境

```
adb shell
```

在运行 Android 的模拟器之后，执行上面的命令，即可进入设备或模拟器的 Linux Shell 环境，在这个环境中，可以执行各种 Linux 的命令。

例如，打开 Windows 的命令行窗口，输入命令 adb shell，再输入 Linux 的进入到/sdcard 目录命令和显示文件列表命令，可显示当前目录下的所有文件列表，如附图 B.1 所示。

4）从模拟器（或设备）上传或下载文件

用户可以使用 adb pull、adb push 命令将文件复制到一个模拟器/设备实例的数据文件或是从数据文件中复制。install 命令只将一个 apk 文件复制到一个特定的位置，与其不同的是，pull 和 push 命令可让用户复制任意目录和文件到一个模拟器/设备实例的任何位置。

- 从模拟器（或设备）中下载文件到主机端，使用以下命令：

```
adb pull <模拟器的文件> <主机目录路径>
```

- 将文件从主机端上传到模拟器（或设备），使用以下命令：

```
adb push <主机端的文件> <模拟器目录>
```

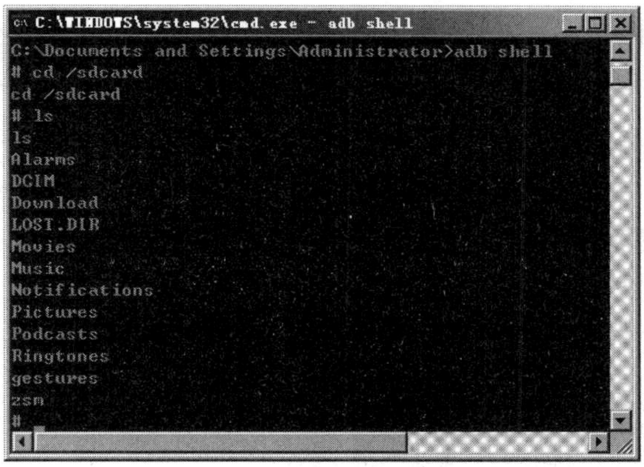

附图 B.1　Linux 的 adb shell 命令

例如，将文件 a.jpg 上传到 SD 存储卡，其命令如下：

```
D:\>adb push a.jpg  /sdcard/a.jpg
182 KB/s (20415 bytes in 0.109s)
```

如附图 B.2 所示。

附图 B.2　从主机 D 驱动器根目录上传文件到模拟器 sdcard 目录

5）启动和关闭 adb 服务器

通常在启动 Android 服务时，adb 服务器也会随之一起启动。当由于某种原因 adb 服务器没有启动时，可以用"adb start-server"来启动。如果不想继续进行调试，可以用"adb kell-server"来关闭 adb 服务器。

2. Android 的 DDMS 工具的使用

DDMS 是 Android 系统的重要工具，DDMS 的全称为 Dalvik Debug Monitor Service。在 Android 应用程序的设计过程中，DDMS 可以为用户提供很多帮助。例如：对程序运行结果截屏，查看正在运行的线程以及堆信息、Logcat 信息、广播状态信息，模拟电话呼叫，接收 SMS，虚拟地理坐标等。

启动 DDMS 工具有两种方式：
- 使用 Eclipse 开发环境所提供的 DDMS 工具；
- 执行 Android SDK 系统安装所在目录下的 tools 中的 ddms.bat 批处理文件。

当打开 Eclipse 后，在开发环境的右上角可以看见 DDMS 图标，单击 DDMS 图标，即可启动 DDMS 工具，如附图 B.3 所示。

如果在 Eclipse 开发环境的右上角没有看见 DDMS 图标，可以在菜单栏中选择"窗口"→"打开透视图"→"其他"→DDMS 命令，启动 DDMS 工具。

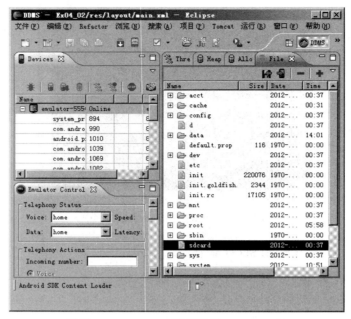

附图 B.3　Eclipse 开发环境的 DDMS 工具

如果要返回 Eclipse 的编辑环境，单击 DDMS 图标旁边的 Java 图标即可。

附录 C　　Map API Key 的申请过程

1. 创建 KML 地图文件

具体方法略。

2. 获取 Map API 密钥

1）找到 debug.keystore 文件所在的目录

其通常位于 C:\Documents and Settings\Administrator\.android 目录中。如果不在此目录中，可以在 Eclipse 中打开"窗口"→"首选项"命令，弹出"首选项"对话框，在左侧的选项栏中选择 Android 下的 Build 选项，在右侧的 Build 面板中查看 debug.keystore 文件所在的目录，如附图 C.1 所示。

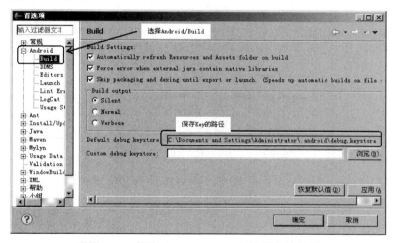

附图 C.1　查看 debug.keystore 文件所在的目录

2）获取 MD5 指纹

打开 CMD 命令窗口，使用 Java 的 JDK 自带的 KeyTool 工具生成 MD5 指纹。在 CMD 命令窗口中输入以下命令：

```
keytool -list -alias androiddebugkey -keystore "C:\Documents and Settings\Administrator\.android\debug.keystore" -storepass android -keypass android
```

系统会提示输入 keystore 密码，此时输入密码（如 android），系统就会输出用户需要的 MD5 认证指纹，如附图 C.2 所示。

3）向 Google 申请 Android Map 的 API Key

打开浏览器，输入网址 http://code.google.com/intl/zh-CN/android/maps-api-signup.html，进入如附图 C.3 所示的页面。

附图 C.2 获取 MD5 指纹

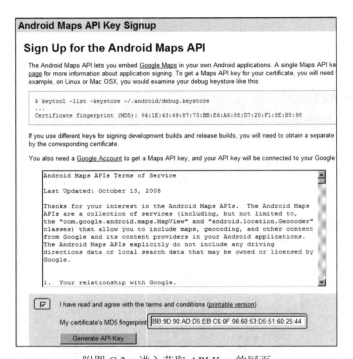

附图 C.3 进入获取 API Key 的网页

单击网页上的 Generate API Key 按钮，用户可以得到需要的 API Key，如附图 C.4 所示。在设计 Google 地图服务的应用程序时，要将密钥添加到界面布局文件 activity_main.xml 中（参见例 10-1）。

附图 C.4 获取 Map API 密钥

图 书 资 源 支 持

感谢您一直以来对清华版图书的支持和爱护。为了配合本书的使用,本书提供配套的资源,有需求的读者请扫描下方的"书圈"微信公众号二维码,在图书专区下载,也可以拨打电话或发送电子邮件咨询。

如果您在使用本书的过程中遇到了什么问题,或者有相关图书出版计划,也请您发邮件告诉我们,以便我们更好地为您服务。

我们的联系方式:

地　　址: 北京海淀区双清路学研大厦 A 座 707

邮　　编: 100084

电　　话: 010-62770175-4604

资源下载: http://www.tup.com.cn

电子邮件: weijj@tup.tsinghua.edu.cn

QQ: 883604(请写明您的单位和姓名)

用微信扫一扫右边的二维码,即可关注清华大学出版社公众号"书圈"。

书 圈